MATHEMATICAL IDEAS AND SOLUTIONS TO UNSOLVED PROBLEMS

VINCE FLYNT

MATHEMATICAL IDEAS AND SOLUTIONS TO UNSOLVED PROBLEMS

PREFACE

Mathematics is considered the most objective science compared to the other sciences such as physics, biology or economics. But it is not without its drawbacks for those who are its practitioners, i.e., mathematicians, being human, could interpret mathematical ideas in their own personal, subjective ways. There could be disputes about mathematical ideas wherein egos could be at play. In other words, there is also disagreement or conflict in mathematics, like any other human activity.

This book will delve into the subject in an open manner and bring up the various approaches to the subject, in particular the various approaches to the solutions to some important unsolved mathematical problems.

Vince Flynt, Ph.D.

CONTENTS

1 MATHEMATICS AND PROOF

Whether mathematical proofs and arguments are "rigorous" enough to be accepted by the mathematical community is a matter of opinion and culture. How do we measure objectively or quantify the "rigor" of a mathematical proof? As a matter of fact, how ought a mathematical proof be like? Take a hypothetical example here. Would mathematical thought exist in the mind of an extra-terrestrial? How would an extra-terrestrial's mind view our Mathematics? Would mathematical proofs make any sense to him?

In the middle of the nineteenth century, the famous logician and philosopher, C. S. Pierce, announced that "mathematics is the science of making necessary conclusions". Mathematics could be "about" anything as long as it is a subject that exhibits the pattern of assumption-deduction-conclusion. The ideal mathematician rests his faith on rigorous proof. He believes that the difference between a correct proof and an incorrect one is an unmistakable and decisive difference. He could think of no condemnation more damning than to say of a fellow mathematician, "He does not even know what a proof is". Yet, ironically, he is unable to supply a coherent explanation of what is meant by rigor, or, what is required to make a proof rigorous. In his own work, the line between complete and incomplete proof is always somewhat fuzzy, and often controversial. Cantor's radical ideas on infinite sets had been ridiculed and he had to suffer nervous breakdowns for it. Now they are regarded as "correct" mathematics. This is not surprising. But, it shows that the subject of "mathematical rigor" itself is controversial, even opinionative and prejudicial. It depends also on the status of the person who puts forward the mathematical proof.

According to an international authority in Engineering Science, who often had to apply mathematics in his work, Professor William F. Taylor, mathematical truth is reasoning that leads to correct physical relationships, where empirical demonstrations are possible. Belief in a non-material reality removes the paradox from the problem of mathematical existence, whether in the mind of God or in some more abstract and less personalized mode. If there is a realm of non-material reality, then there is no difficulty in accepting the reality of mathematical objects which are simply one particular kind of non-material object. Hence, a mathematical proof could be regarded as a procedure by which a proposition about the unseen reality could be established with finality and accepted by all adherents. It could be observed that if a mathematical question has a definite answer, then different mathematicians, using different methods, working in different centuries, would find the same answers. A well-known mathematician even ventured to call mathematics a form of religion. To him, mathematics ought to be regarded as the true religion.

In the opinion of some, the name of the mathematics game is proof - no proof, no mathematics. In the opinion of others, this is nonsense - there are many games in mathematics. Mathematics, then, is the subject in which there are proofs. Traditionally, proof was first met in Euclid. Millions of hours have been spent in class after class, in country after country, in generation after generation, proving and reproving the theorems in Euclid. After the introduction of the "new math" in the mid-nineteen fifties, proof spread to other high school mathematics such as algebra, and subjects such as set theory were deliberately introduced so as to be a vehicle for the axiomatic method and proof. In college, a typical lecture in advanced mathematics, especially a lecture given by an instructor with "pure" interests, consists entirely of definition, theorem, proof, definition, theorem, proof, in solemn and unrelieved concatenation. Why is this so? If, as claimed, proof is validation and certification, then one might think that once a proof has been accepted by a competent group of scholars, the rest of the scholarly world would be glad to take their word for it and to go on. Why do mathematicians and their students find it worthwhile to prove again and yet again the Pythagorean theorem or the theorems of Lebesque, Wiener or Kolmogoroff? Proof serves many purposes simultaneously. In being exposed to the scrutiny and judgment of a new audience, the proof is subject to a constant process of criticism and revalidation. Errors, ambiguities and misunderstandings are cleared up by constant exposure. In its best instances, a proof increases understanding by revealing the heart of the matter. It also suggests new mathematics. The novice who studies proofs, even incomplete proofs, gets closer to the creation of new mathematics. A proof is ritual and a celebration of the power of pure reason. Such an exercise in reassurance may be very necessary in view of all the messes that clear thinking clearly gets us into.

The logical analysis of mathematics, which reduces a proof to an (in principle) mechanizable procedure, is a hypothetical possibility, which is never realized in full. Mathematics is a human activity, and the formal-logical account of mathematics is only a fiction - mathematics itself is to be found in the actual practice of mathematicians. An interesting phenomenon should be noted in connection with the difficulties of proof comprehension. A mathematical theorem is called "deep" if its proof is difficult. Some of the elements that contribute to depth are non-intuitiveness of statement or of argument, novelty of ideas, complexity or length of proof-material measured from some origin which itself is not deep. The opposite of deep is "trivial" - this word is often used in the sense of a put-down. But, it does not follow that what is trivial is uninteresting, useless or unimportant. Despite this hierarchical ordering, what is deep is in a sense undesirable, for there is a constant effort towards simplification, towards the finding of alternative ways of looking at the matter which trivializes what is deep. We all feel better when we have moved from the analytic or abstract toward the analog or practical portion of the experiential spectrum.

According to Imre Lakatos, in his Proofs And Refutations, mathematics, too, like the natural sciences, is fallible, not indubitable. It too grows by the criticism and correction of theories which are never entirely free of ambiguity or the possibility of error or oversight. Starting from a problem or a conjecture, there is a simultaneous search for proofs and counter-examples. New proofs explain old counter-examples. New counter-examples undermine old proofs. To Lakatos, "proof" in this context of informal mathematics does not mean a mechanical procedure which carries truth in an unbreakable chain from assumptions to conclusions. Rather, it means explanations, justifications, elaboration, which make the conjecture more plausible, more convincing, while it is being made more detailed and accurate under the pressure of counter-examples. Each step of the proof is itself subject to criticism, which may be mere skepticism or may be the production of a counter-example to a particular argument. A counter-example which challenges one step in the argument is called by Lakatos a "local counter-example", while a counter-example which challenges, not the argument, but the conclusion itself, is called a "global counter-example". Proofs And Refutations is an overwhelming work. The effect of its polemical brilliance, its complexity of argument and self-conscious sophistication, and, its sheer weight of historical learning all dazzle the reader. In fact, Lakatos had been idolized.

On the one hand, we have real mathematics, with proofs which are established by "consensus of the qualified". A real proof is not checkable by a machine or computer, or even by any mathematician not privy to the gestalt, the mode of thought of the particular field of mathematics in which the proof is located. Even to the "qualified reader", there are normally differences of opinion as to whether a real proof (that is, one which is actually spoken or written down) is complete or correct. These doubts are resolved by communication and explanation, never by transcribing the proof into first-order predicate calculus. Once a proof is "accepted", the results of the proof are regarded as true (with very high probability). It may take generations to detect an error in a proof. If a theorem is widely known and used, its proof frequently studied, if alternative proofs are invented, if it has known applications and generalizations and is analogous to known results in related areas, then it comes to be regarded as "rock bottom". In this sense, all of arithmetic and Euclidean geometry are "rock bottom".

On the other hand, to be distinguished from real mathematics, we have "meta-mathematics" or "first-order logic". As an activity, this is indeed part of real mathematics. But, as to its content, it shows a structure of proofs which are indeed infallible "in principle". We are thus able to study mathematically the consequences of an imagined ability to construct infallible proofs. We could, e.g., give constructivist variations on the rules of proof and see what the consequences of such variations are.

There are various schools of mathematical thought as well. For example, there are the logicist and the constructivist schools of thought. If mathematics were indeed neutral and objective, why is this so? Even the foundation of mathematics, which is set theory, was found to be shaky by Godel, whose Incompleteness Theorems caused deep cracks in the important monumental works of such mathematicians as Frege and Russell. Is mathematics a part of nature that is waiting to be discovered or is it an invention of the human mind? The author would consider mathematics a bit of both, for mathematical proofs are subject to the interpretation and imagination of the human mind. For instance, how could we find an example of $\sqrt{-1}$ in reality (What is $\sqrt{-1}$ actually equal to?), though it is commonly used in mathematics?

In conclusion, the author would like to state that a mathematical proof ought to be closely scrutinized and ought to be regarded as true (only to a high probability). It ought to lead mathematicians into thinking more deeply into the problem and possibly coming up with a better or more interesting proof or proofs. Most mathematical papers which are published in the journals are actually regurgitation and clarification of established mathematical proofs. Mathematical papers with original and fresh ideas that arouse considerable interest and thought are rather rare. Whether a mathematical proof is indeed correct or not is a matter of disputation. Take the twin primes problem for instance where some of the best brains in mathematics have been trying without success to prove the infinity of the twin primes for centuries. There is perhaps one way to indubitably convince a top-flight mathematician of the infinity of the twin primes. That is, if God Himself brings him before a huge screen upon which all the twin primes are infinitely projected in running and ascending order of magnitude. Not surprisingly, some mathematicians may expect a cleverer proof than this. How should the proof be then?

2 MORE ON MATHEMATICS AND PROOF

A mathematical proof could be controversial and might have some not so obvious loop-holes. A proof is information which shows that a statement or a proposition is true. The quantitative aspects of mathematics and reasoning are actually self-evident and require no further proofs than the fact that their self-evidence could be simply checked or confirmed by counting, weighing or measuring. (Is it really necessary to resort to set theory, which is qualitative, in-concrete, not so obvious, abstract and rather subjective evidence, to prove the consistency or logical soundness of arithmetic?) It is the qualitative part, which is intangible, incapable of being counted, weighed or measured and lacking in obviousness, that requires something more concrete (more obvious) as evidence of its soundness or correctness.

The concept of infinity is often used in mathematics. Infinity is an abstraction and cannot be quantified. Since this is so, the proof of the infinity of something, e.g., the primes, would be inherently qualitative and, therefore, controversial. This is perhaps the reason why mathematical problems involving infinity are difficult.

Mathematicians expect mathematical proofs to be infused with rigorous logic. But, it is known that logic is not always consistent as paradoxes have been discovered (by Godel). A person might analyze the premises of a mathematical, scientific, commercial or sociological reasoning. The contents of the premises might vary for these different subjects, but they all might have a common denominator which makes the premises relate with each other, leading to a valid conclusion. This common denominator is "logic". But, the problem is that it is difficult to ascertain the "logic" of abstract principles, whereas the "logic" of quantities (which are tangible) or, arithmetic, could be easily confirmed or checked by counting, measuring, weighing or comparing. Our mind is able to straightaway see "logic" in the above-mentioned subjects if the reasoning involves quantifiable objects. Grasping the logical links between qualitative concepts is an apparently more arduous task. Since quantitative reasoning is obvious, indubitable, it is curious that mathematicians would want to use axioms and set theory to prove the consistency of arithmetic (logical correctness concerning quantities). It is truly the validity or correctness of qualitative reasoning (reasoning with unquantifiable, abstract objects, such as infinity) that is hard to ascertain, for how does one measure, weigh, count or compare unquantifiable objects or qualities?

The Intuitionist school of thought, championed by Dutch mathematician, Brouwer, had wanted to dispense with the law of the excluded middle, wherein a proposition is either right or wrong (though Brouwer had finally been coerced into accepting its validity for finite sets only). For a finite sequence, e.g., one to ten (1, 2, 3, 4, 5, 6, 7, 8, 9, 10), if it is not one, then it is any number from two to ten. For an infinite sequence (such as: 1, 2, 3, 4, 5, 6, 7, 8, 9, 10, …), it is not so straightforward to infer thus. However, for qualities such as tall and short, or, right and wrong, there could be a "middle" state where a thing or person is neither tall nor short, nor, right nor wrong - a person could be of average height and an action could be deemed right under a set of circumstances and wrong under another. Therefore, the principle of "reductio ad absurdum" - that is, to prove that a proposition is true by finding a contradiction in the assumption that it is false - is not consistent. A proposition that is not false might not be true either, i.e., it could be neither true nor false. To conclude, e.g., that a man who is not clever is certainly stupid, is stupid in itself. What if the man is of average intelligence?

The logic of quantities is obvious (as they could be checked or confirmed by counting, weighing, measuring or comparing). The problem lies with counting, measuring or weighing qualities, if it could be carried out at all. The following equation, which displays "quantitative" logic, might be modified to display "qualitative" logic, as follows:-

$$1 + 1 = 2 \quad \text{("quantitative" logic)} \tag{1}$$

$$1 \text{ apple} + 1 \text{ pear} = 2 \text{ apples, if } 1 \text{ apple} = 1 \text{ pear} \tag{2}$$

$$\text{or, } a + b = 2a, \text{ if } a = b \tag{3}$$

The "quantitative" logic of Equation (1) is self-evident or obvious. The difficulty of "qualitative" logic is represented by the following queries or doubts:-

In what ways is the one apple above the same as or equal to the one pear, or, in what ways is $a = b$?

1) Is it the weight?
2) Is it the dimensions?
3) Is it the color?
4) Is it the surface texture?
5) Is it the taste, and so on?

There is indeed no apparent or obvious similarity between the apple and the pear, or, between a and b - their similarities were just assumed - in fact, their similarities, if any, would be controversial.

Godel's proof pertaining to the incompleteness of a formal system of logic or the logic of arithmetic, which is not completely accepted by mathematicians, is a proof by contradiction. It is far from obvious and very difficult to comprehend. Its reasoning is "qualitative". Euclid's proof of the infinity of the primes falls into the same mold; it is as follows:-

Assume P is the largest existing prime in the following sequence of primes:

2, 3, 5, 7, 11, 13, 17, 19, 23, …P

There would always be a prime, Q, or, a prime factor of Q, that is larger than P, whatever the value of P is, comprising the composite of all the primes from 2 up to P, plus 1, which is as follows:

$$Q = (2 \times 3 \times 5 \times 7 \times 11 \times 13 \times 17 \times \ldots \times P) + 1$$

This contradicts the proposition that P is the largest existing prime number, and, therefore, the number of primes is infinite.

There is a quirk in the above-mentioned sequence of prime numbers. Whereas all the other prime numbers are each an odd number divisible only by 1 and itself, 2 is the only prime number which is even and divisible only by 1 and itself. Probably, it ought to be replaced by 1, which, perhaps, could be regarded as an odd number that is divisible only by itself but (probably for the reason that it is also the divisor of all the numbers) it is not regarded as a prime number. This may look arbitrary. Euclid's proof is proof by "reductio ad absurdum" and is hence inconsistent, and, controversial, according to the Intuitionist school of thought (which posits that a proposition is either correct, incorrect or un-decidable).

The application of logical reasoning could sometimes lead to some preposterous results regarding infinity. An example is Zeno's famous paradox regarding Achilles racing the tortoise, whereby it is posited that Achilles would never overtake the tortoise as there would always be an "infinity of intervals" between them. In actual fact, Achilles on account of his much greater speed would always be able to overtake the tortoise, outrunning the latter effortlessly. Infinity is either an invention or discovery of the human mind. It is a controversial, mind-blowing concept, for we are all used to finite systems.

The solving of the Four-Color problem in 1976 by Haken and Appel was not without some controversy though its validity has now been largely accepted by the mathematical community. Haken and Appel solved the problem by reducing it to a "building block" configuration of 1,482 maps whereby proving that these 1,482 "building block" maps are four-colorable would imply that all maps are four-colorable. They worked on these 1,482 "building block" maps with powerful computers for 1,200 hours of computing time. They showed that these 1,482 "building block" maps are indeed four-colorable thus proving that all maps are four-colorable. The snag here is that mathematical proofs have been normally checked by humans. In this case it was not possible for a human to check the proof. What if the computer had indeed made a computing error? There was no way for anyone to check and confirm that the computer had not made any computing error and this had to be accepted on faith. This proof was the first of its kind and was a precedent for the mathematical community.

Though e and pi represent transcendental numbers, we might never be certain that their transcendence might not end at some point towards infinity. But for practical purposes we accept their transcendence.

If the Intuitionist school is right mathematical proofs embracing the law of the excluded middle might not be acceptable. We have yet to come to terms with our understanding of mathematical proofs. "Quantitative" mathematics or arithmetic is obvious, self-evident - we could easily count, weigh, measure or compare quantities - we therefore need not offer a more elaborate proof for an arithmetic. Likewise we do not need to prove our existence by logical inference, for it is obvious or self-evident that we exist. (No question about it. A person just needs to look at the reflection of himself in the mirror and would find that his reflection would have a one-to-one correspondence with his actual self.) But French mathematician and philosopher, Descartes, inferred his own existence by the simple argument "I think, therefore I exist." He, and Godel as well, had resorted to a mythical kind of logical reasoning to prove the existence of God. If man could actually see God, does he still need to resort to logic to prove His existence? As they say "Seeing is believing". In a court of law witnesses are normally called to give evidence, witnesses who have seen or witnessed something related to the case. Why don't the judges use pure logical reasoning instead of calling witnesses to help determine a case, if logical reasoning were so reliable? In mathematics it should be what is not obvious that requires a formal proof and it is actually the "qualitative" or "judgmental" aspects of mathematics that are not so obvious.

What is mathematics really? There is no complete agreement on what mathematics actually is. Some mathematicians and philosophers regard it as the invention of the human mind while others attribute it to human discovery (which means that mathematics exists even if the human mind does not exist). Whichever it is, no one could be certain. However, the author thinks it is a combination of both. It is therefore important that one keeps an open mind on all mathematical ideas.

3 GODEL'S PROOF AND ITS EFFECT ON MATHEMATICS

According to Godel's proof, which has had far-reaching effects on logic and mathematics, any formal axiomatic system, e.g., arithmetic, contains un-decidable propositions; in other words, it contains sentences S such that neither S nor the negation of S can be proved. This is Godel's first incomplete theorem. Its corollary, Godel's second incompleteness theorem, states that a formal system's consistency, e.g., that of arithmetic, cannot be proved by means using the formalization of the system itself; the proof can only be derived by using a stronger system. In short, some statements or propositions cannot be proved, e.g., the statement, "This statement is false".

What Godel meant was that the un-decidable propositions are the result of an inherent weakness in the axioms of mathematics - a formal system's consistency can only be proved by using a more powerful set of axioms. This implies that the axioms of mathematics should be replaced by a stronger set of axioms if there were to be no inconsistency in mathematics. The question is whether there would be mathematicians or logicians clever enough to conceive such a strong set of axioms. In the words of computer programming, if we input garbage the computer would churn out garbage - garbage in, garbage out. Mathematicians therefore ought to be cautious of what they put into mathematics if they do not want garbage results.

The author would like to raise a few points of curiosity here. If, by a quirk of fate, if Earth were to be remade and human beings were to be reborn, would mathematics, mathematical proof and logic ever be the same? Are mathematics, mathematical proof and logic accidents of nature or fate? If civilizations or life exist in other galaxies, what would mathematics, mathematical proof and logic be like there?

Apparently because we human beings have ten fingers and could count with our ten fingers our number system is "tens-based", e.g., 10; 100; 1,000; 10,000; 100,000; 1,000,000; et al. If we had, e.g., 13 fingers, 20 fingers, or, 35 fingers, imagine what our number system would have become, and, what would our mathematics, mathematical proof and logic then be? Can you imagine a "thirteens-based", "twenties-based", or, "thirty-fives-based" number system? What would then be the odd numbers and the even numbers? Would the concepts of "oddness" and "evenness" then have to be modified? Is the nature of mathematics, mathematical proof and logic pre-determined by the physical form and nature of the human being and the form and nature of his environment? If human beings had been unable to count, if they had no fingers, or, if they had no sight, would there ever be mathematics (which could be regarded as the science of numbers), and, if mathematics were still able to "co-exist" with human beings, what would it be like? Yet, many mathematicians and logicians would view mathematics and logic as a reality which is independent of the existence of the human being. However, on this last point, the author harbors some doubt; the author wonders whether mathematics could have existed at all if all human beings had been born without fingers and eyes and had been unable to count!

And, yet, there have been savants, even idiot savants, who have been able to compute very large numbers in their heads at unbelievable speeds. This seems to have been some miracle of nature. Therefore, couldn't there be savants (or geniuses) who could perform similarly impressive and miraculous feats of mathematical reasoning or logical reasoning? Such a person, if he had existed, could have out-thought or out-reasoned a genius such as Godel.

We could only hope that such a savant or genius would materialize to reform mathematics and mathematical reasoning so that there would be no more inconsistencies in mathematics and logic, so that there would really be a solid foundation for mathematics and logic. This genius could perhaps re-invent mathematics and produce many great theorems and axioms.

4 GODEL'S UNDECIDABLE THEOREM AND MATHEMATICS

Godel postulated that there are statements or propositions that could neither be proved nor disproved to be true and created a sort of revolution in mathematics, i.e., not all the formalism of mathematics could be proved or disproved rigorously - there are always some paradoxes to bewilder the mathematician.

The following is one such paradox:-

a) The writing on one side of a sign-board says, "The statement on the other side of this board is false".
b) The writing on the other side of the sign-board says, "The statement on the other side of this board is true".

If (a) above is a true statement, (b) contradicts it, and, if (b) is a true statement, (a) contradicts it.

The following statement is also paradoxical:-

This statement is false.

If this statement is false, this statement is true, and, if this statement is true then this statement is false.

Then, there is also the paradox concerning classes or sets. For example, the class of all classes is a class. Similarly, the class of all catalogues, to take another example, is a catalogue. But, here goes the paradox: The class of all dogs is not a dog.

Zeno's Paradox is another well-known paradox. This paradox concerns a race between Achilles and a tortoise. The tortoise, being a slow creature, is given a head-start over Achilles. But, Achilles would never be able to catch up with the tortoise, let alone overtake it. If we keep on halving the distance between Achilles and the tortoise to indicate how Achilles would soon catch up with the much slower tortoise, we would find that even if we go on halving the distance between Achilles and the tortoise till infinity the tortoise would always be ahead of Achilles though the distance between the tortoise and Achilles would get smaller and smaller with each halving. Thus, Achilles would never be able to catch up with the tortoise, let alone overtake it. This runs counter to common sense and experience. We know that Achilles, moving at a greater speed than the tortoise, would overtake the tortoise at some point in time. We have no doubt about this. We would be able to observe this happening. But, by the above-mentioned logic, this is impossible. Here, the author would like to pose the question: Should we trust logic or our physical senses more?

There is little doubt that the mathematician or logician is likely to trust logic more than the non-mathematician or non-logician, for logic is the tool he uses to create new mathematics and abstractions. The man in the street might leave logic to the ivory tower logician and be content with just being able to see Achilles overtaking the tortoise.

Mathematics is a subject that is concerned with proofs or evidences that certain mathematical statements are true. But Godel's undecidable theorem states that such proofs, or, dis-proofs, are not achievable all the time. At the metaphysical level, or, more fundamental level, we might not even be able to prove our very own existence or actuality (Could one accept Descartes' proof of his own existence, which is as follows: I think, therefore I am?), our sanity or the reality of our senses; sometimes, we could not even tell whether an event or occurrence had been a dream or had been real. There also seems to be some inconsistency in mathematical proofs. Mathematical proofs rely on axioms, which are obvious, unproven statements, lemmas, which are proven statements leading to a more important statement or statements, and, theorems, which are proven statements. Mathematicians demand rigor in mathematical proofs, but they could accept unproven statements which are obvious to them, i.e., axioms. However, what appears as obviously true to one mathematician might not be obviously true to another mathematician. Furthermore, proofs acceptable to one group of mathematicians might not be acceptable to another group, e.g., the proof by contradiction which is commonly used in mathematical proofs is not acceptable to the group who call themselves intuitionists. The twin primes conjecture, which postulates that there is an infinitude of twin primes or primes separated by 2, e.g., had not been proven or disproved so far despite many attempts by mathematicians, though practically all mathematicians believe that the conjecture is true. The twin primes conjecture may be one of those statements in mathematics which are true but whose truth is not provable, in accordance with Godel's undecidable theorem. Since thousands, millions, or billions, of twin primes (this quantitative evidence, though it might be acceptable as proof of the infinitude of the twin primes by scientists, is not acceptable to the mathematicians) have been found and the conjecture is so obviously or apparently true though its (complete) proof is lacking (perhaps, the proof or disproof is impossible, vide Godel's undecidable theorem), the twin primes conjecture could perhaps be treated as an axiom instead, i.e., it could be regarded as an obvious but unproven statement. Perhaps, to avoid such inconsistency, mathematics should do away with axioms, which may not be possible. For example, Boolean algebra has provided the axioms for a rigorous application of mathematical abstraction; it gives us the "and", "or" and "not" statements, which are fundamental in logic, and whose equivalent in mathematics are "times", "add" and "subtract".

With the inconsistency and paradoxes encountered it is little wonder that mathematicians feel rather nervous about their field of work. Godel's theorems have cast a doubt on the soundness of mathematical logic, to the dismay of many mathematicians who swore by it, e.g., the great mathematician, David Hilbert.

5 ON GODEL'S INCOMPLETENESS THEOREM

How should we think or act in a logical manner? We should exercise pain, caution and care when making statements, e.g., do our research first and get our facts or premises right, make a careful choice of words, terms or expressions to be used, aim at clarity and at being understood, listen to, consider and accept or adopt others' ideas or points of view if they are relevant, win the support of or acceptance by others for our logical propositions or ideas, et al. A logical statement could be simple and short or it could be complex, detailed and lengthy. The facts or premises contained in the statement should be true facts, facts whose truthfulness could be verified. It is important that the statement could be verified or proved to be true, e.g., confirmed by an experiment or experiments, or, some other kind of test. For those statements whose truths are not certain or verifiable, we could make them with some qualification or caveat - we could make such statements with a probabilistic, but reasoned, approach, e.g., we could state that something is probably true, most probably true, unlikely to be true, has little likelihood of being true, in all probability true (or untrue or false), true under certain circumstances, or, false under certain circumstances, et al., depending on our intuition and how strongly we felt about the probability of its being true, or, false, though we were not certain of or able to verify its truth or falseness. It is of course important to have clarity and alertness of mind while making statements.

The following is a listing of the kinds of statement we may encounter from our fellow-beings or counterparts:-

1) The statement is entirely true. (The truth is verifiable.)
2) The statement is entirely false. (The falseness is verifiable.)
3) The statement is partially true and partially false. (The partial truth and partial falseness are verifiable.)
4) The statement is partially true and partially false, while the rest of the statement is not verifiable as true, or, false but may be considered probably, or, most probably true, or, false. (The partial truth and partial falseness are verifiable.)
5) The statement is partially true, while the rest of the statement is not verifiable as true, or, false but may be considered probably, or, most probably true, or, false. (The partial truth is verifiable.)
6) The statement is partially false, while the rest of the statement is not verifiable as true, or, false but may be considered probably, or, most probably true, or, false. (The partial falseness is verifiable.)
7) The statement is neither true nor false, but may be considered probably, or, most probably true, or, false. (The statement could not be verified to be true, or, false.)
8) The statement is entirely true under certain circumstances. (This is verifiable.)
9) The statement is entirely false under certain circumstances. (This is verifiable.)
10) The statement is partially true and partially false under certain circumstances. (These partial truth and

partial falseness are verifiable.)

11) The statement is partially true and partially false under certain circumstances, while the rest of the statement is not verifiable as true, or, false but may be considered probably, or, most probably true, or, false. (These partial truth and partial falseness are verifiable.)

12) The statement is partially true under certain circumstances, while the rest of the statement is not verifiable as true, or, false but may be considered probably, or, most probably true, or, false. (This partial truth is verifiable.)

13) The statement is partially false under certain circumstances, while the rest of the statement is not verifiable as true, or, false but may be considered probably, or, most probably true, or, false. (This partial falseness is verifiable.)

14) The statement is neither true nor false under certain circumstances, but may be considered probably, or, most probably true, or, false. (This statement could not be verified to be true, or, false.)

15) The statement is neither true nor false under certain circumstances, and there is little or hardly any probability that it is true, or, false. (This statement could not be verified to be true, or, false.)

16) The statement is none of the above, i.e., it is neither true nor false under any circumstances, nor probably, or, most probably true, or, false - we cannot make anything or form any conclusion about the statement at all. (This statement could not be verified to be true, or, false.)

The above listing of 16 kinds of statement which we may encounter may not be an exhaustive listing. The listing may be further classified, e.g., there may be a statement which is three-quarter true and one-quarter false, a statement which is two-third true and one-third probably, or, most probably true, or, false, a statement which is one-third true, one-third false and one-third probably, or, most probably true, or, false, a statement which is one-fifth true, two-fifth false and two-fifth probably, or, most probably true, or, false, et al. There could be lots of fine, subtle distinctions in the logical statements or propositions, which we should be keenly aware of. We should be alert to all the possible implications.

Recall that by Godel's incompleteness theorem there are true statements whose truth is not provable and false statements whose falseness is not provable. The question is if we could not prove or verify the truth or falseness of a statement how could we be certain or know that the statement is true, or, false? (Is it just a hunch, feeling or intuition?) Wouldn't it be a contradiction (of mathematical reasoning wherein rigorous or solid proof is demanded) or absurd to state thus "This statement is true (or, false) but its trueness (or, falseness) cannot be proved"? The most we could say about such a statement is that it is probably, or, most probably true, or, false. Godel appears to have had erred in this respect, for it is practically not possible to know whether a statement or proposition is true, or, false, if its truth, or, falseness is not verifiable or provable, i.e., a statement could only be true, or, false, if it could be proved to be so - otherwise, it would be just a conjecture, e.g., a mathematical statement which has not been proved is a conjecture while a mathematical statement which has been proved is a theorem. In mathematics, a statement is true only if it is proved to be so. Could an "undecidable" or "unprovable" mathematical statement be true as per Godel's incompleteness theorem (which is actually a contradiction of the important mathematical principle of the need for proofs)? In other words, could we say "By Godel's incompleteness theorem, this mathematical statement is true but its proof is an impossibility", or, "I know this mathematical statement is true though its proof is an impossibility"? Which could invite a counter-argument "If you cannot prove that this mathematical statement is true, how do you know that this mathematical statement is true?", which would make it all look rather absurd.

6 NUMBERS

Numbers are meant to be finite, counted, and for describing quantities. Though numbers are finite, they do lead to infinity, i.e., if one were to start counting numbers one could count to no end if one wanted to. Bear in mind that quantity is a fixed amount. Hence, any number must be a fixed amount.

Numbers take various forms; they take the form of rational or irrational numbers, prime, even, or odd numbers, et al. These symbolize the property, or, "quality" of numbers. Their charm is something that appeals to the intellect and the imagination.

Numbers have no beginning and no end. They need not begin with "1" or end with, e.g., "1,000,000". They can begin with "minus something", e.g., -10, and end with something towards infinity. Numbers can form series or sequences of definite order or indefinite order. Patterns are discernible in numbers. Series or sequences of numbers can be of ascending order or descending order, arithmetic progression or geometric progression. For example, $3 + 6 + 9 + 12 + 15 + 18 + 21 + 24 + 27 \ldots$ is an infinite series with an arithmetic progression and $1 + 1/2 + 1/4 + 1/8 + 1/16 + 1/32 + 1/64 + 1/128 + 1/256 \ldots$ is an infinite series with a geometric progression, and, 3, 6, 9, 12, 15, 18, 21, 24, 27, ... is an infinite sequence whose numbers increase in an arithmetic order and 1, 2, 4, 8, 16, 32, 64, 128, 256, ... is an infinite sequence with a geometric progression.

Numbers are rational if they are whole numbers or integers, or if they take the forms of fractions, e.g., 3/8, 7/9, 11/201, et al. Only rational numbers can be classified as "even" or "odd". Even numbers are sums of two even numbers or two odd numbers. Odd numbers are indivisible by even numbers such as 2, 4, 10, et al., except by odd numbers, such as 3, 5, 7, et al.; they are each a composition of an odd and an even number added together, e.g., 5 is actually $3 + 2$, and, 9, $7 + 2$, where 2 is even and 3 and 7 are odd numbers. An irrational number may be an algebraic number (as are all rational numbers) or a transcendental number (e.g., e). A real number is either a rational number or an irrational number. A complex number is the sum of a real number and an imaginary number, the latter being equal to the product of a real number with i (i represents the square root of -1). Cardinal and ordinal numbers are the more abstract entities pertaining to sets.

Prime numbers are odd numbers with a special quality, viz., divisibility by itself and by 1 only, and nothing else, e.g., 41 is only divisible by 1 and 41, which gives results of 41 and 1 respectively. The prime numbers are also regarded as the "atoms" or building-blocks of the integers or whole numbers.

The study of operating with numbers is known as arithmetic. However, in higher mathematics, e.g., calculus, symbols are also important. Numbers represent quantities and they can signify big or small quantities. They, together with symbols, are the objects of logical reasoning in mathematics. Arithmetic merely involves addition, subtraction, multiplication and division, but in higher mathematics substitution involving symbols comes in. For example, in calculus, symbols such as "∂" or "Δ" stands for change in rate or variance, "∞" stands for infinity and "$\sum$" stands for summation, and, in algebra or geometry, alphabets such as, e.g., "x", "y", "a" and "b" are used to denote unknown quantities or distances.

In mathematics numbers and alphabets are closely akin. For example, the infinite sequences, 1, 2, 3, 4, 5, 6, 7, 8, … and a, b, c, d, e, f, g, h, …, each represents a certain order, a consecutive order. "1" can be regarded as equivalent to "a", "2" to "b", "3" to "c" and so on, indefinitely (one-to-one correspondence). The theory of sets deals with such equivalences. In set theory, which was created by Georg Cantor, a set within which is the sequence, 1, 2, 3, 4, 5, 6, 7, 8, …, and a set within which is found the other sequence, a, b, c, d, e, f, g, h, …, are both equivalent sets. Set theory is studied in modern mathematics and is a fundamental theory in the theory of numbers. In the above case, it is possible to substitute for "1" the letter "a", for both are equivalent. If, however, 1 + 2 = 3, then a + b = c, an equivalence. Intersection sets, and joint and disjoint sets are other sets which are studied in set theory. Intersection sets give us a clearer understanding of subtraction, as shown below:-

(a)   (b)

      a, b, c   --   a, c   =   a   b   c

As shown above, (b) is the result of intersection, or actually, subtraction.

Joint sets are shown below:-

(a)   (b)

      a, b, c   a, b, c   =   a, b, c

The two sets in (a) are joint together to give (b). This gives the impression of equality or sameness for indeed the letters, a, b, c , in the two sets in (a) are the same. Disjoint sets give the opposite impression, the impression of inequality or difference.

The study of the theory of numbers cannot be complete without the study of set theory. In the sciences, e.g., physics, electronics and chemistry, numbers play an important role, though such subjects are mainly concerned with practical theories. But some of these theories could be worked out mathematically; that is when numbers come into play. In theoretical physics, e.g., technical symbols incorporated with numbers make the subject highly abstruse, which requires knowledge and experience to tackle. In calculus, in differentiation, the quantity without a root at the end of the equation, e.g., + 7, or, - 36, is omitted, and the coefficients and roots in the equation are varied in value and power respectively; in any case, the coefficient is increased in value and the root is decreased in power, e.g., $2x^8$ (2 here is the coefficient), on differentiation, becomes $16x^7$ (the coefficient now becomes 16, obtained by multiplying 2, the coefficient, by 8, the power of the root, x, and the root, x, is decreased in power from 8 to 7, by 1); in some cases, the root is omitted, as in, e.g., 5x (x here is the root) which becomes 5. In integration, the effect is the reversed of differentiation; the equation would be lengthened, unlike in differentiation. These are some of the rules in calculus which mathematicians have to abide by.

Differentiation is used to calculate variation in something, e.g., rate of increase, or, decrease, in speed, while integration is used to calculate such things as area and volume.

Symbols which precede numbers, e.g., the "+" symbol which signifies "plus" and the "-" symbol which signifies "minus", are important. The "+" symbol is normally left out, if functionless; e.g., we know when we look at the number "8" that it means "+8" as well. But the "-" symbol would always be denoted. In calculus, in the computation of distance or variance, a "+" sign before a number signifies a positive value and a "-" sign before a number signifies that the number has a negative value. For example, if the distance from A to B, being positive, is 8 miles (or, +8 miles), the negative value, which is the distance measured from B to A, would be -8 miles, the positive and negative signs in this case symbolize opposite directions of measurement. In vector calculus, e.g., the plus and minus signs indicate opposite directions, such as the directions of fluid flows.

If one looks at the diagram below, one would understand this point:-

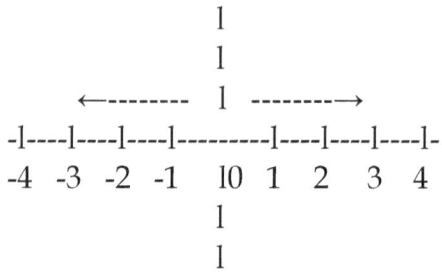

Notice the direction of increase of the positive values, 1, 2, 3, 4, and the direction of increase of the negative values, -1, -2, -3, -4. They are opposite, the positive values increase to the right, as is traditional, the negative values increase to the left.

In the diagrams below, A is assumed to be 0 in (1) and B is to the right and is 10 miles (or, +10 miles) away from 0 or A. In (2) B is now assumed to be 0 and A is to the left of B or 0. As any value to the left of 0 is normally regarded as negative, the distance of A from B is to be regarded as -10 miles. In this case, -10 miles are 10 miles, in terms of distance. Only the directions of measurement and starting points of measurement are different. But strictly speaking, i.e., arithmetically speaking, 10 cannot be equal to -10. 10 can only be equal to -10 if the conditions stated above are present.

(1)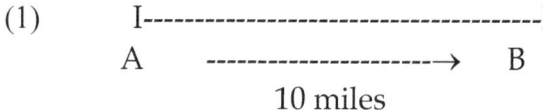

(2)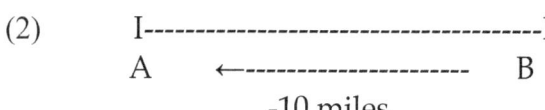

Numbers and symbols cannot be dissociated from one another; any number is understood to have either a positive or a negative value, hence, a positive or a negative symbol. All numbers which are not negative are understood to have a positive or "+" sign preceding them; it is unnecessary to put the "+" sign before the numbers with positive values unless addition or summation is involved.

The theory of numbers like any other theory evolves from other systems. It should thus be treated in relation to the systems of mathematics evolved by the various great mathematicians such as Isaac Newton, Georg Cantor, Jules Henri Poincare, Georg Friedrich Bernhard Riemann, Niels Henrik Abel and Evariste Galois.

7 EQUATIONS

What is an equation? $A = B$. $C = y + x$. $S = x^2 + 2yx + y^2$. These are equations. The first two equations are linear equations. The third equation, where x and y each has a power of 2, is a nonlinear equation. Notice the "is equal to" symbol (i.e., =). A, e.g., is still equal to B after an increment in value provided that B's value is also increased by a margin equal to that of A, e.g., x; hence, $A + x = B + x$. On the other hand, $A - x = B - x$. Notice that when you decrease the value of A you have to decrease the value of B as well by the same margin, x, in order that A in its new form equals B.

Look at this equation, $A = B = C$. Is it not obvious that $A + B = B + C$ is true? Is it not true then that $A + C = 2B$, $A + B = 2C$, $B + C = 2A$? Look at the proof. Substitute for A, B, since B is equal to A, and C, also B, since B is also equal to C. Hence, $B + B = 2B$. Hence, $A + C = 2B$ is true. Equations are meant to be solved. Simple equations can be solved by practically anyone who is acquainted with algebra. It is perhaps the quadratic equations which present the difficulty. The roots (variables or unknowns, typified, e.g., by the symbols x and y) of the equations however can be found by anyone who is acquainted with the methods of finding the roots of equations and who has gone through the woes of solving equations.

Now, equations are of different degrees. There are, e.g., first-degree, second-degree, third-degree and fourth-degree equations (the degree of an algebraic equation or polynomial is the highest power or sum of powers in any term of the equation, e.g., $x^4 + 2x^2 - x$ and xy^2z are both of the fourth degree). It is possible to solve the first-degree, second-degree, third-degree and fourth-degree equations, but not considered possible to solve the fifth-degree (quintic) or higher degree equations in general. One might be able to solve any of these equations by finding the square or cube root of the whole equation, or, if two (or more) equations were given, one could multiply one or both of the equations by a certain number or numbers and subtract one equation from the other, or, add both equations, so as to bring about, finally, a less complex equation with only one root (where there is only one symbol, e.g., x, or, y, which is one of the roots of the equation), which could hence normally be solved without much difficulty.

Let us go further. An equation could be expressed, e.g., as "the sum/subtraction of something and something and something is equal to something", e.g., $x^2 + 2xy - 4y^2 = 6$, where x and y, the roots of the equation, could each be seen to have the power of two. To find the roots or in other words the values of these symbols (i.e., x and y) is the problem of algebraists. In many cases, a few equations are given, the values of the roots of each of these equations must be assumed to be the same. Look at the following equations (we may call them simultaneous equations):-

$$x^2 + 4x^2y + y^2 = 8 \qquad (1)$$

$$2x^2 + 8x^2y + 4y^2 = 30 \qquad (2)$$

The way we should go about finding the roots of the above equations is to multiply one of the equations by a certain number or both the equations by two different numbers (each equation by one of the two numbers), and, after that, either subtract one equation from the other or add up both the equations so as to bring about a new equation with only one root (or symbol/variable/unknown) in the form as follows: $2y^2 = 14$ (therefore: $y =$

$\sqrt{14} \div 2 = \sqrt{7}$). Once this reduction of the two equations to a simple equation with only one root has been carried out, the problem of solving the equations is almost complete. Let us carry out this operation here. Let us multiply Equation (1) above by 2 (($x^2 + 4x^2y + y^2 = 8$) x 2 = $2x^2 + 8x^2y + 2y^2 = 16$) to get the following equation:-

$$2x^2 + 8x^2y + 2y^2 = 16 \qquad (3)$$

Now, subtract Equation (3) above from Equation (2). After this operation we get the following equation with only one root (y):-

$$2y^2 = 14 \qquad (4)$$

Therefore: $y = \sqrt{14} \div 2 = \sqrt{7}$

In this way, we obtain the value of the root, y, which is $\sqrt{7}$. After finding the value of one root (in the above case, y) of the equations (Equation (1) and Equation (2) above), the value of the other root (in the above case, x) of the equations (Equation (1) and Equation (2) above) could be found by substituting the root (y, in the above case) in one (anyone) of the two equations (Equation (1) or Equation (2) above) with its newly found value ($\sqrt{7}$, in the above case) and carrying out the necessary computation. If we, e.g., substitute y in Equation (1) above with $\sqrt{7}$, we would get the following:-

$$x^2 + 4x^2\sqrt{7} + 7 = 8$$

$$x^2 + (4 \text{ x } \sqrt{7})x^2 = 8 - 7 = 1$$

$$x^2(1 + (4 \text{ x } \sqrt{7})) = 1$$

Therefore: $x = \sqrt{1} \div (1 + (4 \times \sqrt{7}))$

Boolean Algebra makes use of the theory of sets. The equations consist of sets of symbols (normally in the form of alphabets). Look at the following:-

$A(B + C) = A(C + B)$

$A(B + C) = AB + AC$

$BA + CA = (B + C)A$

Notice the inter-positioning of letters.

In any case, one of the axioms concerning such algebra is that the product of two symbols, e.g., A and B, in a certain order is not necessarily equal to the product of the two same symbols in another order. AB is not necessarily equal to BA. In this case it should be borne in mind that A stands for anything, except a number, e.g., an action, such as acting (A being the initial for acting), and B for anything, except a number, e.g., boating. Why is AB not necessarily equal to BA? See the following explanation:-

Eating comes before digestion, for we have to take in food before digestion, which is the decomposition of food into simpler and more soluble elements, can occur. It is therefore absurd to say that digestion comes before eating. Let eating = E and digestion = D. So it could be seen that ED = DE is not true, for ED is in the right order as it represents the order that eating must come before digestion, while DE is not so, as it represents the reverse. Such equations could rightly be called equations of orders; one order is not necessarily equal to another. To solve or understand such equations, the symbols should be carefully noted and understood as representing whatever things they represent, and their orders should be realized and understood as either right or wrong - a right order cannot be equal to a wrong order.

In geometry, AD = DA is correct. AD represents the distance between points A and D measured from A and DA represents the same distance measured from D.

The symbols cannot be the symbols representing numbers, for if "A" represents the number "1" and "D" the number "2", then AD = DA is incorrect for then it means 12 = 21 which is obviously not true. The integer, 12, is never quantitatively equal to the integer, 21, for 21 is quantitatively greater than 12 by 9. But if by 12 = 21 one means that 12 is qualitatively equal to 21 then one may be right; 12 may be equal to 21 in that they are comprised of the same integers (1 and 2), i.e., integers quantitatively the same.

Hence, in the study of equations one should find out in what sense something is equal to something. Is it quantitative equality or qualitative equality? Normally in equations the symbols so used stand for numerical values; therefore, quantitative equality is involved. But this is not so in other forms of algebra, such as Boolean Algebra, where symbols may stand for anything apart from numbers. In Boolean Algebra, the symbols may stand for certain classes of things. The axiom is such that the class of an object belonging to a certain class of objects is equal to the class of this certain class of objects. Equations in different branches of mathematics should be differently understood and dealt with. In the new mathematics, which involves the theory of sets, in geometry, and in elementary algebra, the equations should be dealt with accordingly. Thus, we could separate equations into classes or types. Algebraic equations and equations of that branch of mathematics known as New Mathematics, or Boolean Algebra (which could be regarded as the algebra of classes), should not be confused, and should not be solved in the same manner.

The study of equations comes under Algebra, which is the branch of elementary mathematics that makes use of symbols representing unknown quantities or variables in order to determine their values by the elementary operations of arithmetic. Equations may involve numerical values, properties of objects, or order of events, things. The study of order itself could in a way be regarded as the study of the properties of numbers. For example, the numbers or integers, 1, 2, 3, 4, 5, 6, 7, …, in this infinite sequence increase in a certain order known as a consecutive order, wherein each consecutive number is always of a value greater than that of the number just before by 1. The numbers of a sequence can increase, e.g., in an arithmetic order or geometric order. 3, 6, 9, 12, 15, 18, 21, … is an infinite sequence whose numbers increase in an arithmetic order with a marginal difference of 3. The marginal differences can be turned into another infinite sequence, 3, 3, 3, 3, 3, 3, …. 1, 2, 4, 8, 16, 32, 64, … is an infinite sequence with a geometric order or progression. The marginal differences can be turned into the infinite sequence, 1, 2, 4, 8, 16, 32, …. The numbers of a sequence can also increase without any order or pattern, e.g., 1, 7, 21, 68, 73, 104, 175, …, which is an infinite sequence with numbers that increase without any order, whose marginal differences can be turned into another infinite sequence, 6, 14, 47, 5, 31, 71, …, whose numbers increase/decrease without any order.

Equations could, hence, be described as the logical aggregation of symbols and numbers in a certain pattern or order. To juggle with these symbols and numbers and so create new equations which are logical is the main task of the mathematician doing work with equations.

8 A DEEPER LOOK AT ALGEBRA

All of us are positive that 1 + 1 = 2 or 2 + 3 = 5. But, has anyone ever thought why this must always be so?

In the author's opinion, 1 + 1 = 2 or 2 + 3 = 5 are, at best, only approximations in the abstract. What exactly do these two equations mean? How can we interpret these two equations? We can interpret them (or misinterpret them) in at least several ways.

Firstly, we ask ourselves 1 (what?) + 1 (what?) = 2 (what?) and 2 (what?) + 3 (what?) = 5 (what?).

Secondly, we may interpret the above equations as 1 (pole) + 1 (pole) = 2 (poles) and 2 (poles) + 3 (poles) = 5 (poles), for example. Hence, we are saying 1 + 1 = 2 and 2 + 3 = 5.

Thirdly, we may construe that 1 + 1 is not necessarily equal to 2 or that 2 + 3 is not necessarily equal to 5. "How come?", you may wonder. Look at the following illustrations:-

a) Can't 1 (2 ft. long pole) + 1 (1 ft. long pole) = 3 (1 ft. long poles), i.e., 1 + 1 = 3, for example?
b) Can't 2 (2 ft. long poles) + 3 (1 ft. long poles) = 7 (1 ft. long poles), i.e., 2 + 3 = 7, for example?

Fourthly, can 1 (apple) + 1 (pear) = 2 (apples), or, 2 (apples) + 3 (pears) = 5 (pears), for example? Apparently, these two equations will hardly hold water now.

We can "justify" the equations 1 + 1 = 2 and 2 + 3 = 5 by ensuring that the numerals such as 1 and 2 in equation 1 + 1 = 2, and, 2, 3 and 5 in equation 2 + 3 = 5, share a "common property". For example, 1 (gram) + 1 (gram) = 2 (grams) , and hence 1 + 1 = 2 ("gram" being the "common property"); 2 (2 ft. long poles) + 3 (2 ft. long poles) = 5 (2 ft. long poles), and hence, 2 + 3 = 5 ("2 ft. long poles" being the "common property").

But, can we be absolutely certain, e.g., that 1 (apple) + 1 (apple) = 2 (apples) , "apple" here being the "common property"? In other words, could 1 (small apple) + 1 (small apple) = 2 (big apples) ? This evidently could not be so.

Let us look at the following equation concerning apples:-

1 (small apple) + 1 (big apple) = 2 (big apples)

Is the above equation valid? This equation can or cannot hold water, depending on whether some "common property" exists or not. Let us look at the following illustrations:-

a) If, e.g., 1 (small apple) + 1 (big apple) = 3 (grams of apple) and 2 big apples = 6 (grams of apple) then, of course, this equation will not be valid (for 3 grams ≠ 6 grams).
b) Similarly, e.g., if 1 (small apple) + 1 (big apple) = 7 (cubic centimeters of apple) and 2 (big apples) = 14 (cubic centimeters of apple), this equation will not be justifiable (for 7 cubic centimeters ≠ 14 cubic centimeters).

This equation can be justified when the following conditions are present:-

a) If, e.g., 1 (small apple) + 1 (big apple) = 4 (grams of apple) and 2 (big apples) = 4 (grams of apple), in which case the "4 grams of apple" is the "common property" belonging to each side of the equation, i.e., when we in effect have "4 grams of apple = 4 grams of apple" (here, common sense will tell us that the big apple on the left side of the equation is heavier than each of the two big apples on the other side of the equation).
b) Also, if, e.g., 1 (small apple) + 1 (big apple) = 7 (cubic centimeters of apple) and 2 (big apples) = 7 (cubic centimeters of apple), in which case the "7 cubic centimeters of apple" is the "common property" belonging to each side of the equation, i.e., when we in effect have "7 cubic centimeters of apple = 7 cubic centimeters of apple" (here, common sense will again tell us that the big apple on the left side of the equation is of greater volume than each of the two big apples on the other side of the equation.

In (a) above 1 (small apple) + 1 (big apple) = 2 (big apples) in terms of weight (grams), while in (b) above they are equal in terms of volume (cubic centimeters).

Thus, "common property" here should be a form of "precise description", e.g., precise measurements such as weight, length, et al. For example, "4 grams" and "7 cubic centimeters" above are "precise properties". Things can only be considered equal if they display similar measurable, even discernable, properties or qualities such as weight, volume and length, and, perhaps, *physical features or appearance (*here, we should specify that things are equal because they are "equal" or similar in physical features or appearance - physical features or appearance tend to be subjective and difficult to judge, unlike the measurable properties such as weight and volume; physical features or appearance would give rise to sets, classes, types or species, hence, the algebra of sets or classes (set theory), for example).

Imprecise descriptions, which cannot be precisely measured or discerned, will not do. For example, 1 (unhappy man) + 1 (unhappy man) = 2 (unhappy men) may not be justifiable, for we cannot measure happiness or ascertain the degree of happiness in each of the men.

So, will, e.g., one plus one always equal two, or, two plus three always equal five, et al.? It depends on how we interpret these algebraic equations. There should be more precision (of interpretation) to algebraic equations. Otherwise, algebraic equations may not really make much sense if we look deeply into them, as illustrated above.

9 INFINITY IN MATHEMATICS

What is infinity? Is it a number, a quantity? Perhaps, we could define infinity as an unquantifiable number, a rather self-contradictory definition. The lay person practically does not bother about infinity as he is apparently concerned about the now and the not too distant future. After all, no one could live forever or to infinity. Scientists have even predicted that life on earth would end some time in the future, the result of some terrible catastrophe. So, what is the point of being so concerned about infinity?

Infinity is apparently an invention of the intellect, an abstraction. Everyone would have no problem understanding what infinity is all about and would realize that it is something which would always be physically and empirically unattainable; one would certainly be never able to prove infinity by any physical experiment, though one might think that one could prove infinity by logic as has been the case of mathematics wherein mathematicians have been proving or trying to prove the infinitudes of, e.g., the primes, the twin primes (as in the twin primes conjecture), the even numbers greater than the number 2 which are each the sum of two prime numbers (as in the Goldbach conjecture) and the zeros of the Riemann zeta function (as in the Riemann hypothesis), et al. We could of course conceive of an infinite outer space but could we be absolutely sure that outer space is infinite, a space that extends straight away from us (and not in a circular way) forever, or, to infinity? For all we know, outer space might be just a circle or circular entity, or, a Mobius-strip-like space, where we could go round forever in circles, as a number of scientists has conjectured, a concept which is more in line with our empirical experiences and which appears more acceptable or credible. However, as the saying goes, fact could be stranger than fiction. Immortality or being able to live forever, e.g., is an imaginable concept and could exist in fiction, e.g., in a fairy tale, but everyone in his right mind knows that it would never be attainable, hence, nobody ever tries to prove that immortality is possible, achievable or attainable. When mathematicians prove or try to prove that the prime numbers, twin primes, even numbers greater than 2 which are each the sum of two prime numbers and the zeros of the Riemann zeta function, et al., are infinite they are apparently implying that the numbers concerned, which are the objects of their study, are real and have a "life" of their own. This is apparently disputable, and all mathematicians apparently do not have a consensus on this. If the above-mentioned numbers, which are the objects of study of a number of mathematicians, might be just inventions of the mathematicians' minds, then should they not be studied or considered only with a "pinch of salt"?

To prove that these numbers, the objects of study of a number of mathematicians, could go on forever, ad infinitum, it is apparently necessary to prove that the space (physical space) in which these numbers would occupy (e.g., when being listed, i.e., put on paper, such as a computer printout) is infinite (mental space could never be infinite as nobody could live forever and count forever). However, this space (which contains the listing, or, computer printout) would extend into outer space. Should we not therefore try to prove that this space, outer space, is infinite, i.e., this space goes on forever, which would allow the above-mentioned numbers, which are the objects of study of a number of mathematicians, to go on forever as well? But, our understanding of outer space appears limited. We apparently do not know what is actually out there in outer space and what it is actually like. We could of course conceive, or, imagine it to be infinite, but its infinity is neither really proven nor a certainty. An infinite outer space would imply infinite possibilities, including things that are apparently impossible. Perhaps, at some point towards infinity in outer space our physical laws and logic would not apply at all. Perhaps, at this point and beyond, life, consciousness and logic could not and would not exist and something or many things or an infinite number of things unimaginable, miraculous, or incredible, would exist or occur.

It is probably easier to prove the existence of God than the infinitudes of the above-mentioned mathematical objects, the various kinds of numbers. In fact, two reputable mathematicians in the past, viz., Rene Descartes and Kurt Godel, had produced proofs or reasoning which supported the existence of God. God as a sentient Being (man is also a sentient being) might be able to provide evidences of His existence, e.g., through the workings of physical phenomena, and might even be able to make His presence physically felt, e.g., someone might claim to have seen God or heard the voice of God. As a sentient Being, God could be expected to have His existence confirmed or proved in this manner. But the infinitudes of mathematical objects such as those described above belong to a different genre. An infinitude is not a sentient being who could be touched, heard or seen with the eyes, but an abstract noun, an abstract object, an object that might not be real at all but is just the result of the human imagination. For example, we might use our imagination to think of something absurd, e.g., we might imagine people living without food and drink, people living without breathing, people living forever, things falling towards the sky on their own without outside force, et al. Since we could conceive such absurdities, should we regard them as realities? Obviously, no. Such things have never happened in life and would apparently never happen in life. Similarly, no one has ever lived to, counted things to or carry out activities to infinity, though everyone could conceive infinity. Therefore, infinitude or infinity should not be considered real, or taken so seriously (only imaginary or conceived, though neither "could be experienced", really knowable nor, thus, really provable, though practically all mathematicians appear to think it is provable).

The author recalls the responses of some editors of some of the mathematical journals to which he has submitted his article on the solutions for the twin primes conjecture. In the article the author has shown several "constructions" or ways of producing an infinite list of twin primes. But these several editors commented that these "constructions" would not work at some point towards infinity, without giving any counter-example or other reason to prove their point. Of course, such a postulation based apparently on gut feeling and not concrete evidence or sound reasoning, such as a counter-example, is unfair and unacceptable. However, in accordance with the arguments presented above, there seem to be some grains of truth in what these editors said. It seems that mathematical problems such as the twin primes conjecture and the Goldbach conjecture are meaningless and really have no sensible solutions, i.e., problems involving infinity, such as the above-mentioned mathematical problems, are meaningless and not really solvable at all (and any so-called solutions found would probably be illusions only). This implies that practically everyone involved with such problems might have wasted their effort and time. In our working life we deal with all kinds of numbers and quantities but there are surely some finite limits to how large these numbers or quantities would be, which would depend on the capacity of our computers. A number which is extremely large, e.g., a number with millions of digits, would be impossible to name and would be meaningless, and would thus have practically no utility - it would be just a curiosity. It would appear more reasonable, realistic, practical, and, useful to carry out the search for the above-mentioned mathematical objects, e.g., the twin primes and the zeros of the Riemann zeta function, up to only a certain realistic but sufficiently large or gargantuan numerical limit, e.g., numbers with some millions, billions, trillions or zillions of digits only, computing facilities permitting (which would of course result in a checkable proof, a proof which could be checked and confirmed with a computer, though many mathematicians of the older school might consider this kind of proof inelegant). There is no practical use in being concerned with what happens beyond this limit, this point, towards infinity, where at some point our laws of nature and logic might not apply anymore. It would probably be better (not to mention easier too) to prove the existence of a Supreme Being; if He really exists the human race could really implore Him to help or bless them. The human race certainly could not expect any help from the infinite prime numbers, the infinite twin primes, the infinite even numbers which are each the sum of two prime numbers, the infinite zeros of the Riemann zeta function, et al., and, solving such mathematical problems would apparently contribute practically nothing at all to the well-being of the human race.

Just understanding the nature of infinity should be good enough. However, the great mathematician Georg Cantor had another interpretation for infinity which has apparently been so bizarre that he had been greatly criticized by a number of his peers, though it is now generally accepted by mathematicians. According to Cantor, there are different levels or sizes of infinity, which could be denoted by the respective cardinal numbers. This seems to be paradoxical, as it appears to be against common sense. Could we then not utilize Cantor's interpretation of infinity to conjecture that there are different infinities of prime numbers, twin primes, even numbers which are each the sum of two prime numbers, zeros of the Riemann zeta function, et al.? (Cantor's interpretation of infinity might be construed to imply this, which would complicate matters.) To Cantor, the infinite set of all real numbers (algebraic plus transcendental) has a higher cardinal number than the infinite set of positive integers, i.e., the infinity of the infinite set of real numbers is larger than the infinity of the infinite set of positive integers. What could be the rationale behind this? The author's interpretation of this rationale is as follows. Assume that x is a point at infinity for both the infinite set of real numbers and the infinite set of positive integers. Assume that a is the number of real numbers between 0 and x in the infinite set of real numbers. Assume that b is the number of positive integers between 0 and x in the infinite set of positive integers. Then, a > b, which means that the set of real numbers between 0 and x is larger than the set of positive integers between 0 and x. If x = ∞ (infinity), then the infinity of the infinite set of real numbers is larger than the infinity of the infinite set of positive integers. Is this reasoning really sound, though on the surface it looks plausible? The author here ponders about a corollary. Consider a house, y (infinite set y). There are chairs c (real numbers) and tables t (positive integers). Does it mean that the house (y) with p number of chairs (c) is larger than the same house (y) with q number of tables (t) when p > q? It does not make sense anymore here. Similarly, Cantor's interpretation of infinity appears unsound.

As infinity should not be considered real (as explained above), mathematical problems having to do with it, such as the twin primes conjecture, the Goldbach conjecture and the Riemann hypothesis, are not realistic and are hence meaningless. A suggestion here is to replace the "infinity" of such mathematical problems with a "realistic numerical limit approaching infinity", a realistic but sufficiently large or gargantuan numerical limit which is countable, quantifiable or computable (and large enough to reasonably represent infinity). This "realistic numerical limit approaching infinity" could be regarded as a "point at infinity" or one of the elements, components or atoms of infinity, and could hence be regarded as representing infinity itself. There is actually no point in having a numerical limit or "point at infinity" which is so forbiddingly large that it is incomprehensible, meaningless and useless. Numbers or magnitudes range from the infinitesimal or infinitely small to the infinite or infinitely large, i.e., from - ∞ to + ∞ (∞ being the symbol for infinity). This could be represented by a diagram showing the various magnitudes or graduations, e.g., infinitely small, very small, small, medium, large, very large, infinitely large, as follows:-

infinitely small (- ∞)/very small/small/medium/large/very large/infinitely large (+ ∞)

Our "realistic numerical limit approaching infinity" or "point at infinity" could be any of the numbers within the box designated "infinitely large", e.g., a number immediately after the boundary between "very large" and "infinitely large"; the "infinitely large" box could be regarded as representing infinity, or, infinity itself, and each and everyone of the numbers within the "infinitely large" box could be regarded as an element, component or atom of infinity. This way of looking at infinity or treating infinity would make infinity a realistic and meaningful concept, and the mathematical problems associated with infinity would become more tractable. This actually represents another way of interpreting infinity and might appear self-delusive. Perhaps it is better for mathematics to do away with the problems associated with infinity, problems such as the twin primes conjecture, the Goldbach conjecture and the Riemann hypothesis, and modify such problems to problems dealing with the quantifiable or computable, problems which deal with the more concrete and more practical, problems which deal with the realistic, wherein finding an x number, a sufficiently large or gargantuan number, of twin primes, even numbers which are each the sum of two prime numbers or zeros of the Riemann zeta function should be sufficient.

Finally, even if we could live to infinity we could never reach infinity because infinity implies that it is inherently unreachable (without end or limit), i.e., if infinity is reachable it is not infinity but a limit. All this implies that infinity is probably not real and not realizable at all if it is real. That is, if it is real there would really be no way to prove or tell that there is indeed no end or limit to the infinite progression. To say that one is able to prove infinity is to say that one is able to achieve it, reach it, experience it, know it indeed or empirically exists or realize it, whence it would not really be infinity (infinity implies the opposite of all of these), a contradiction or paradox. Proving infinity by some other means, e.g., reasoning by "reductio ad absurdum" or contradiction, or, induction, would probably be just illusory, without any empirical or really solid basis, perhaps just a play of ideas (infinity, unlike other quantities, is not something which could be counted, measured, weighed or easily visualized or conceived, but is a very highly abstract, even mind-boggling, entity); both the proof by contradiction and the proof by induction, though they have been generally accepted mathematical proofs, are apparently not "foolproof" and without critics (e.g., those belonging to the Intuitionist school of mathematics have raised objections to the proof by contradiction), having some apparent weaknesses (as explained elsewhere in this tome). This means that infinity is probably or likely to be just imaginary, an invention of the mind, a curiosity. This means that infinity should not be taken so seriously.

10 AXIOMATIC METHOD

There is an evident inherent inconsistency or arbitrariness in the axiomatic method in mathematics.

Axioms, being obviously or inevitably true statements (without any need for a proof), may be a necessity in order for a mathematical reasoning to proceed; axioms are the premises based on which the mathematical reasoning proceeds, e.g., the axioms in Euclid's THE ELEMENTS. That is, without axioms, or, premises, the reasoning cannot be carried out - there is nothing to reason with.

However, we should be mindful of the use of axioms while carrying out our mathematical reasoning, as axioms may be arbitrary. What is an axiom or obviously true statement to one may not be so to another. For example, "1 + 1 = 2" is obviously true to all and can be regarded as an axiom (without any need of a proof). And yet in their monumental treatise PRINCIPIA MATHEMATICA Bertrand Russell and Alfred North Whitehead took a couple of hundreds of pages of dense mathematical reasoning to prove this simple, obvious fact.

Some of the great conjectures in mathematics also appear intuitively true, or, obvious, to many but their proofs are still being sought. Can't we regard these conjectures as axioms, being obviously true, once sufficient practical evidences are there? For example, the Riemann Hypothesis, considered the most important unsolved problem in pure mathematics, has been shown to be practically true as many billions of the zeros of the zeta function have been found (it is said that more than one billion of them are discovered everyday by researchers) and is still waiting for a mathematical proof; some researchers are so certain of its correctness that they adopt the Riemann Hypothesis as an axiom in their mathematical reasoning.

Can't obviously true conjectures, e.g., the above-mentioned Riemann Hypothesis, be regarded as axioms instead? What is the criterion for an assertion being acceptable as an axiom, if not for its obviousness or inevitability? There is evidently an inherent inconsistency, arbitrariness or self-contradiction about axioms. Why is it that certain mathematical statements which are obviously true can be accepted as axioms (without the need of a proof) while other mathematical statements just as obvious need a proof? There is also the question of the level of understanding of a person, which varies from individual to individual - what is obvious to an intelligent person may not be so to a less intelligent one, which implies that a person who needs an explanation, or, proof to make a statement obvious to him may be lacking in intelligence. So how do we decide and who decide what statements are obvious and can be regarded as axioms, e.g., the two apparently intelligent, even brilliant, authors of the above-said monumental PRINCIPIA MATHEMATICA evidently could not accept the statement "1 + 1 = 2" as an axiom and needed a few hundred pages of dense mathematical reasoning to affirm the statement's validity (this act could be interpreted as the act of two foolish persons splitting hairs and might also imply that the two were lacking in intelligence)? All this appears arbitrary.

11 A LOOK AT MATHEMATICAL REASONING

The solutions to mathematical problems presented in this book are more or less based on new combinations of old ideas. The author might not have fully agreed with some of these old ideas, e.g., the argument by contradiction and the concepts relating to infinity, for instance, the different orders of infinity, but since they are now "conventional wisdom" which appears acceptable to practically all mathematicians, the author just adopted them as "common grounds for agreement" to get the job done. The author would therefore regard the solutions as being rather subjective as there was the element of the desire to include ideas which he surmised that the referees of any mathematical journals considering the articles for publication would find acceptable; to be too original and to include totally new ideas might only invite suspicion, and, rejection.

Are the referees of the mathematical journals really objective and fair-minded enough to give the solutions a fair consideration? Though the author might be wrong and risks giving the impression of crying "sour grapes", he could not help but feel that most journal referees are overly cautious of cranks and somewhat biased - they appear to generally favor a certain style of mathematical presentation and complexity of argument - they seem to foster the impression that mathematics is difficult and must appear difficult - the editor of one of these mathematical journals even had the affront to reply (even before the author had submitted the article to him) that if the author were not a professional mathematician, he could not be expected to have the correct solution. Is this how a professional mathematician should conduct himself? We should all be sorry for mathematics if many professional mathematicians are like that. There are much more pressing and important problems to be solved than all the very difficult, outstanding mathematical problems. So what if the twin primes conjecture or the Goldbach conjecture is right, or, wrong? Does it make any iota of difference at all to our lives or our well-being? Its only contribution perhaps is that it gives the author and some like-minded people a sense of satisfaction, or, sense of achievement, which is indeed nothing of great importance.

However, mathematics has been effectively used to model nature, and, it has been said that nature is mathematical. Mathematics is not that different from a game, such as chess, or any other game, in that there are rules to be followed. In presenting the solutions to the mathematical problems, the author had to make an attempt to stick to some rules in order not to be "disqualified", rules which might appear arbitrary. It is as though the author was playing a game, e.g., chess.

Besides, the author always had the fear that the journal referees would not understand his ideas, and he had to attempt to present his ideas in as simple and least complicated a manner as possible in order that they be understood and approved by the referees, whoever they might be. Of course, if the author's ideas were not understood, his articles would never stand a chance of publication by the journals.

In the articles on the twin primes conjecture and the Goldbach conjecture, the author has shown a number of ways of creating, producing or generating respectively twin primes, and, even numbers which are each the sum of two primes, all the way to infinity. This represents a constructive proof. On top of this, the author has used the "reductio ad absurdum" proof or proof by contradiction (which, as stated before, is not acceptable to mathematicians who abide by the Intuitionist school of thought and favor the constructive proof). Various other arguments have also been presented. If these were not acceptable as evidence which supports the conjectures, then what kind or kinds of evidence are acceptable? Strangely and apparently illogically, the managing editor of one mathematical journal had replied the author that the methods of producing or generating pairs of twin primes presented in his article would fail at infinity. If the managing editor had proffered a good reason, or, proof, why the methods would fail at infinity, then the author could have been forced to change his stand, but none had been offered. What then constitutes an acceptable mathematical proof or evidence? This appears to be the central problem of mathematics - no mathematician seems to know, and, it seems more like a matter of opinion, subjective.

In mathematics, even if trillions of twin primes, or, even numbers which are each the sum of two primes, could be produced or generated with the help of very powerful computers, this is still not regarded as good enough evidence which supports either conjecture, though to the scientist such overwhelming evidence could be regarded as proof that a hypothesis is correct. The twin primes conjecture and the Goldbach conjecture involve infinity, and, since we could not quantify infinity and might not even have a good, decent definition or understanding of infinity, it would apparently be very difficult to prove the infinity of the twin primes or the even numbers which are each the sum of two primes.

A proof is information which supports a proposition or statement. It is something which certifies that a proposition or statement is correct. For example, how do we prove or certify that Sudan is hot? Could we do it by mere logical deduction? For instance, logical deduction such as the following:-

1) Everyone I spoke to who has been to Sudan complains that it is hot.
2) The geography books I consulted state that Sudan is hot.
3) Therefore, Sudan is indeed hot.

There is certainly a much better way, in fact a scientific way, of proving or certifying that Sudan is hot. We go to Sudan, station ourselves at a representative place there, and monitor the temperatures there with thermometers at certain regular intervals for a specific period. Then we total all the temperature readings taken over this period and average them up. We have first of all to decide on and specify what temperature range is considered "hot climate". We compare the average temperature computed against this temperature range, and, if it falls within this range, we could conclude, or certify, that Sudan is indeed hot.

Next, how do we prove that a person is a thief, or murderer, for example? The chain of reasoning could be as follows:-

1) The accused was not at home, or, at work, at the time the crime was committed.
2) When questioned, he claimed he was drinking at a pub at the time the crime was committed.
3) A check with the pub manager, and, the CCTV recordings showed he was never at the pub, which implies he had lied.
4) He had no other reason to lie except for covering up his crime.
5) We therefore conclude that he was the one who committed the crime.

On the other hand, what if there were other evidences, e.g., a CCTV record or film of the accused committing the act of crime, and, his fingerprints or DNA (if he had injured himself and bled) at the scene of the crime? Wouldn't these be incontrovertible, incriminating evidence or proof of his guilt, as compared to the proof of his guilt by mere logical deduction?

As for the twin primes conjecture and the Goldbach conjecture, the only truly incontrovertible proof of the infinitude of the twin primes and the even numbers which are each the sum of two primes is to be at infinity (if only this were possible) and check or confirm whether the twin primes and such even numbers would still be found. To arrive at a conclusion for these two conjectures by mere logical deduction is at best only a second-guess, the next best method of proof in lieu of actual, experimental proof. Physical or experimental evidence or proof would be more assuring than proof by mere logical reasoning. In other words, physical and concrete evidence or proof should take precedence over abstract proof. However, for the case of the twin primes conjecture and the Goldbach conjecture, since infinity is something not physically achievable at all but is just an abstract (we might call it "invented") entity, there is no other way of proving these conjectures one way or another apart from logical deduction.

The author feels that trying to prove the infinity of anything might be just trying to do the impossible, which is absurd. Having said all this, does it not imply that the author has been a fool in attempting the solutions for the two conjectures? In a sense, the author has been a fool. He could not resist taking up a great intellectual challenge. He could not resist the pleasure of applying clever ideas or ploys in solving these two challenging problems. However, the author thinks that his solutions are reasonable, even clever, or, cunning. And, he would suggest that they be only regarded, at best, as correct, until proven wrong. The author would be very happy if any of the solutions is proved wrong. He would then be happy because of the efficacy of logic, he would be happy for logic.

12 A PROOF OF THE TWIN PRIMES CONJECTURE

Euclid's proof of the infinitude of the primes has generally been regarded as elegant. It is a proof by contradiction, or, *reductio ad absurdum*, and it relies on an algorithm which will always bring in larger and larger primes, an infinite number of them. However, the proof is also subtle and has been misinterpreted by some with one well-known mathematician even remarking that the algorithm might not work for extremely large numbers. This chapter presents a strong argument which supports the validity of the twin primes conjecture, using reasoning similar to that of Euclid's proof of the infinity of the primes.

In 1919, Viggo Brun (1885 - 1978) proved that the sum of the reciprocals of the twin primes converges to Brun's constant:

$$1/3 + 1/5 + 1/7 + 1/11 + 1/13 + 1/17 + 1/19 + ... = 1.9021605 ....$$

It is evident that the twin primes thin out as infinity is approached. The problem of whether there is an infinitude of twin primes is an inherently difficult one to solve, as infinity (normally symbolised by: ∞) is a difficult concept and is against common sense. It is impossible to count, calculate or live to infinity, perhaps with the exception of God. Infinity is a nebulous idea and appears to be only an abstraction devoid of any actual practical meaning. How do we quantify infinity? How big is infinity? We could either attempt to prove that the twin primes are finite, or, infinite. If the twin primes were finite, how could we prove that a particular pair of twin primes is the largest existing pair of twin primes, and, if they were infinite, how could we prove that there are always larger and larger pairs of them? It is evidently difficult to prove either, with the former appearing more difficult to prove as the odds seem against it. This chapter provides proof of the latter, i.e., the infinitude of the twin primes.

Let 3, 5, 7, 11, 13, 17, 19, …….., n - 2, n be the list of consecutive primes, wherein n & n - 2 are assumed to be the largest existing twin primes pair, within the infinite list of the primes.

Let 3 x 5 x 7 x 11 x 13 x 17 x 19 = a .

Lemma: (a x …….. x n - 2 x n) - 2, &, (a x …….. x n - 2 x n) - 4 will never be divisible by any of the consecutive primes in the list: 3, 5, 7, 11, 13, 17, 19, …….., n - 2, n, whether they are prime or composite. (See Appendix 1.)

This implies that:

If (a x …….. x n - 2 x n) - 2 &/V (a x …….. x n - 2 x n) - 4 are prime, then:

(a x …….. x n - 2 x n) - 2 > (a x …….. x n - 2 x n) - 4 > n > n - 2

If (a x …….. x n - 2 x n) - 2 &/V (a x …….. x n - 2 x n) - 4 are non-prime/composite, then:

(a) each prime factor, e.g., y below, of (a x …….. x n - 2 x n) - 2 > n > n - 2
(b) each prime factor, e.g., z below, of (a x …….. x n - 2 x n) - 4 > n > n - 2

(a x …….. x n - 2 x n) - 2 = prime V composite   (1)

(a x …….. x n - 2 x n) - 4 = prime V composite   (2)

(1) & (2) = twin primes, if both (1) & (2) are prime

(1) & (2) > n & n - 2

Let Y represent the prime factors of (a x …….. x n - 2 x n) - 2 if (a x …….. x n - 2 x n) - 2 is not prime (i.e., it is composite), each prime factor may pair up with another prime which differs from it by 2 to form twin primes. Let y = prime factor in Y .

y & y +/- 2 = twin primes, if y +/- 2 is prime

y & y +/- 2 > n & n - 2

Let Z represent the prime factors of (a x …….. x n - 2 x n) - 4 if (a x …….. x n - 2 x n) - 4 is not prime (i.e., it is composite), each prime factor may pair up with another prime which differs from it by 2 to form twin primes. Let z = prime factor in Z .

z & z +/- 2 = twin primes, if z +/- 2 is prime

z & z +/- 2 > n & n - 2

Therefore: (a x …….. x n - 2 x n) - 2 > (a x …….. x n - 2 x n) - 4 > y V y +/- 2 V z V z +/- 2 > n > n - 2

By the above, the following, which implies that n & n - 2 are the largest existing twin primes pair, is an impossibility:

n > n - 2 > (a x …….. x n - 2 x n) - 2 > (a x …….. x n - 2 x n) - 4 > y V y +/- 2 V z V z +/- 2

It is hence clear that no n & n - 2 in any list of consecutive primes can ever possibly be the largest existing twin primes pair and larger twin primes than them can always be found by applying the same mathematical logic (as is described in Appendix 1), e.g., by utilising the evidently effective Algorithm 1, or, Algorithm 2 described in Appendix 3. That is, a largest existing twin primes pair is an impossibility, which implies that the twin primes are infinite. It is possible to find larger twin primes than n & n - 2 no matter how large n & n - 2 are, with the following formulae involving the list of consecutive primes: (a x …….. x n) - 2 & (ax …….. x n) - 4, which by the nature of their composition are capable of generating new primes/twin primes which will always be larger than n & n - 2 (see Appendix 1); this operation is part of Algorithm 1 described in Appendix 3. This is an indirect proof or proof by contradiction (reductio ad absurdum) of the infinity of the twin primes, for our assumption of n & n - 2 as the largest existing twin primes pair will be contradicted by the discovery of larger twin primes, implying the infinity of the twin primes. Again, by applying the same mathematical logic (described in Appendix 1), by way of this evidently effective Algorithm 1 in Appendix 3, and going one step further, we can find that many twin odd integers found between n and (a x …….. x n) - 2, which differ from one another by 2 and are not divisible by any of the primes in the list of consecutive primes: 3, 5, 7, 11, 13, 17,19, …….., n, will be twin primes larger than n & n - 2, our assumed largest existing twin primes pair, which is a contradiction of this assumption, thus implying or proving the infinitude of the twin primes. In this manner, i.e., by resorting to Algorithm 1 in Appendix 3, by continually adding more and more consecutive primes to the list of consecutive primes: 3, 5, 7, 11, 13, 17, 19, …….., n, i.e., continually extending the value of n, and utilising the formula: (a x …….. x n) - 2, as well as the formula: (ax …….. x n) - 4, to perform the computations a la Algorithm 1 in Appendix 3, many larger and larger twin primes can be found, all the way to infinity, in parallel with the infinitude of the list of consecutive primes: 3, 5, 7, 11, 13, 17, 19, …….. of which the twin primes are a part together with other primes pairs, wherein the twin primes are not likely to be finite (as is evident from Appendix 2) and can be expected to be infinite. (Algorithm 2 in Appendix 3 may also be utilised for this purpose but it is evidently a longer and less efficient method.) A largest existing twin primes pair is indeed an impossibility. The twin primes are infinite.

CONCLUDING REMARKS

There are 376 pairs of twin primes (752 primes) found within the 2,500 consecutive primes from 2 to 22,307 - this means that 30.08%, which is sizeable, of the 2,500, not a small quantity, consecutive primes are twin primes. 3, 5 & 7 are the only "triple" primes found. There is no regularity in pattern in the appearance of the twin primes, except that the intervals between consecutive twin primes vary greatly by from 4 integers to 370 integers - the intervals between the consecutive twin primes increase and decrease, and, then increase and decrease again, by turns, giving rise to a graph that is characterised by many peaks, i.e., the curve is rough and nonlinear, making its description (hence, forecast of the twin primes) by differential equations practically impossible.

The argument used here to prove the twin primes' infinity is the indirect (reductio ad absurdum) method, which had been used by Euclid and other mathematicians after him. Logically, 1 or 2 examples of "contradiction" should be sufficient proof of infinity, for it does not make sense to have a need for an infinite number of cases of "contradiction", as our proof would then have to be infinitely and impossibly long, an absurdity. This method of proof is "proof by implication" as a result of "contradiction" - which is a "short-cut" and smart way in proving infinity, instead of "proving infinity by counting to infinity", which is ludicrous, and, impossible. Hence, 1 or 2 cases of "contradiction" should be sufficient for implying that there would be an infinitude of twin primes, which of course also tacitly implies that there would be an infinitude of the number of cases of such "contradiction". (Euclid evidently had this logical point in mind when he formulated the indirect (reductio ad absurdum) proof of the infinity of the primes.) This method of proof had been cleverly used by a number of mathematicians, not the least by the great German mathematician, David Hilbert. For example, Hilbert had used an indirect method (the "reductio ad absurdum" proof) to prove Gordan's Theorem without having to show an actual "construction", a proof which had been accepted by his peers.

The chapter presents 2 algorithms for generating or sieving all the twin primes in any range of odd numbers - by utilising any of these 2 algorithms (preferably the evidently more efficient Algorithm 1), we will be able to find many twin primes which are all larger than those in any chosen list of consecutive primes, i.e., we will be able to generate many larger and larger twin primes. This is indeed significant. There is evidently some deep meaning in the ease with which the twin primes turn up, as is shown in this chapter. It is thus evident that the twin primes are an inherent characteristic of the infinite prime numbers (as well as odd numbers), a characteristic which could be regarded as "self-similar" or "fractal". A twin primes pair is in effect any pair of odd numbers which differ from one another by 2 and are indivisible by any number except itself, the negative of itself, +1 and -1 (i.e., the pair of odd numbers are prime numbers). Any consecutive odd numbers or odd numbers that differ from one another by 2 are therefore potential prime numbers, as well as potential twin primes, and, the likelihood of them being prime is infinite (vide Euclid's proof and Dirichlet's Theorem), i.e., the primes will always be found amongst them and will be there all the way to infinity (the primes being evidently the "atoms" or building-blocks of all the whole numbers or integers, i.e., all the odd numbers and even numbers - every odd number or integer is either a prime number or composite of prime numbers (i.e., the integer has prime factors), and, every even number is the sum of two prime numbers (vide the Goldbach conjecture which, it appears, practically all mathematicians believe to be true), as well as the product of prime numbers (composite)); hence, the likelihood of them being twin primes is infinite as well (the twin primes being an inherent property of the infinite prime numbers - as well as odd numbers - the twin primes can in fact be likened to next-door neighbours, which are a common, expected thing (see Appendix 4)).

So far, there has not been any indication or confirmation that the number of twin primes is finite and the so-called largest existing pair of twin primes has not been found and confirmed (which of course would be impossible to find and confirm if the twin primes were infinite). On the other hand, practically everyone could intuit that the number of twin primes is infinite.

Due to the evident effectiveness of the 2 algorithms described in Appendix 3 in bringing in larger and larger twin primes, the above proof of the infinitude of the twin primes is not only an indirect proof or proof by contradiction (reductio ad absurdum), importantly, it is also a constructive proof. It should be noted that the characteristic of a mountain or infinite volume of sand is reflected in the characteristic of some grains of sand found there so that studying the characteristic of some grains of sand found there is enough for deducing the characteristic of the mountain or infinite volume of sand, to ascertain the quality of a batch of products it is only necessary to inspect some carefully selected samples from that batch of products and not every one of the products and to carry out a population census, i.e., find out the characteristics of a population, it is only necessary to carry out a survey on some carefully selected respondents and not the whole population. With any of these 2 algorithms (preferably the evidently more efficient Algorithm 1), in like manner, by the same principle, we could carry out a study of a carefully selected list of integers and their associated primes and twin primes and deduce by induction whether the twin primes would always turn up, appear infinitely, in the list which is itself infinite - this act is rather like extrapolation.

APPENDIX 1

Note: The (only) even prime 2 is omitted from the list of consecutive primes: 3, 5, 7, 11, 13, 17, 19, ..., n - 2, n stated in the chapter, wherein n & n - 2 are assumed to be the largest existing twin primes pair.

The list of newly created primes, and, twin primes for n = 5, 7, 11, 13, 17, 19, …….. (n = 19 being the maximum limit achievable with a hand-held calculator) is as follows:-

1] For n = 5, we get the following new primes/new twin primes:

$(3 \times 5) - 2 = 13$ (α)
$(3 \times 5) - 4 = 11$ (β)

2] For n = 7, we get the following new primes/new twin primes:

$(3 \times 5 \times 7) - 2 = 103$ (α)

$(3 \times 5 \times 7) - 4 = 101 \quad (\beta)$

3] For n = 11, we get the following new primes/new twin primes:

$(3 \times 5 \times 7 \times 11) - 2 = 1,153 \quad (\alpha)$
$(3 \times 5 \times 7 \times 11) - 4 = 1,151 \quad (\beta)$

4] For n = 13, we get the following new prime and composite number with its prime factors:

$(3 \times 5 \times 7 \times 11 \times 13) - 2 = 15,013 \quad (\alpha)$ - Prime Number
$(3 \times 5 \times 7 \times 11 \times 13) - 4 = 15,011 \quad (\beta)$ - Composite Number (= 17 x 883, with 17 pairing with 19 to form a twin primes pair and 883 pairing with 881 to form another twin primes pair)

5] For n = 17, we get the following new primes/new twin primes:

$(3 \times 5 \times 7 \times 11 \times 13 \times 17) - 2 = 255,253 \quad (\alpha)$
$(3 \times 5 \times 7 \times 11 \times 13 \times 17) - 4 = 255,251 \quad (\beta)$

6] For n = 19, we get the following new prime and composite number with its prime factors:

$(3 \times 5 \times 7 \times 11 \times 13 \times 17 \times 19) - 2 = 4,849,843 \quad (\alpha)$ - Prime Number
$(3 \times 5 \times 7 \times 11 \times 13 \times 17 \times 19) - 4 = 4,849,841 \quad (\beta)$ - Composite Number (= 43 x 112,787, with 43 pairing with 41 to form a twin primes pair while 112,787 is a stand-alone prime)

.
.
.
.

Results Of α And β Above
1) α above generates 6 new primes (13; 103; 1,153; 15,013; 255,253; 4,849,843), nil

composite numbers.

2) β above generates 4 new primes (11; 101; 1,151; 255,251), 2 composite numbers (15,011 = 17 x 883; 4,849,841 = 43 x 112,787).

3) α and β above together produce 4 pairs of new twin primes (13 & 11; 103 & 101; 1,153 & 1,151; 255,253& 255,251).

4) The prime factors of α and β above form 3 pairs of new twin primes with prime partners which differ from them by 2 (19 & 17; 43 & 41; 883 & 881).

5) All the new twin primes in (3) and (4) above are larger than n & n - 2, the assumed largest existing twin primes pair, which is indirect proof of the infinitude of the twin primes.

Why It Is Impossible For Any n & n - 2 To Be The Largest Existing Twin Primes Pair

α = (3 x 5 x 7 x 11 x 13 x 17 x 19 x .... x n) - 2, and, β = (3 x 5 x 7 x 11 x 13 x 17 x 19 x .... x n) - 4 will never be divisible by any of the consecutive prime numbers in the list: 3, 5, 7, 11, 13, 17, 19, ..., n, whether they are prime or composite (non-prime and divisible by prime numbers or prime factors). This means that none of the consecutive prime numbers in the list: 3, 5, 7, 11, 13, 17, 19, ..., n can ever be factors of α and β, and, α and β must be new primes/twin primes larger than all the consecutive prime numbers in the list: 3, 5, 7, 11, 13, 17, 19, ..., n, or, if they were composite (non-prime and divisible by prime numbers or prime factors), their prime factors (and "twin prime" partners which differ from them by 2) must be larger than all the consecutive prime numbers in the list: 3, 5, 7, 11, 13, 17, 19, ..., n. This is a very important mathematical logic, which needs to be grasped in order to understand the proof.

This all implies that no n & n - 2 (if n - 2 were also a prime number) in any list of consecutive prime numbers can ever possibly be the largest existing twin primes pair, since all the new primes/twin primes produced or generated by α and β will always be larger than n & n - 2. That is, a largest existing twin primes pair is an impossibility, which implies the infinitude of the list of the primes/twin primes.

In other words, by the mathematical logic stated above, which explains why all the new primes/twin primes, which α and β by the nature of their composition are capable of producing or generating, will always be larger than n & n - 2, no n & n - 2 in any list of consecutive prime numbers: 3, 5, 7, 11, 13, 17, 19, ..., n can ever possibly be the largest existing twin primes pair, i.e., a largest existing twin primes pair is indeed an impossibility, thus implying the infinitude of the list of the twin primes. This is a very important inference.

Regardless of how long the list of the twin primes pairs is, it is possible to find some new twin primes pairs which will always be larger than n & n - 2, our assumed largest existing twin primes pair - the largest twin primes pair in our assumed finite list of the twin primes pairs, with α and β, which is indirect proof of the infinity of the twin primes. In fact, by the same principle, many twin odd integers found between n and α, which differ from one another by 2 and are not divisible by any of the primes in the list of consecutive primes: 3, 5, 7, 11, 13, 17, 19, ..., n, will be twin primes pairs larger than n & n - 2, our assumed largest existing twin primes pair, which is a contradiction of this assumption, hence implying or proving the infinitude of the twin primes. (Refer to Algorithm 1, as well as Algorithm 2, in Appendix 3.)

APPENDIX 2

Anecdotal Evidence Of The Infinity Of The Twin Primes

TOP TWIN PRIMES IN 2000, 2001, 2007 & 2009

In the year 2000, $4648619711505 \times 2^{60000} \pm 1$ (18,075 digits) had been the top twin primes pair which had been discovered. In the year 2001, it only ranked eighth in the list of top 20 twin primes pairs, with $318032361 \cdot 2^{107001} \pm 1$ (32,220 digits) topping the list. In the year 2007, in the list of top 20 twin primes pairs, $318032361 \cdot 2^{107001} \pm 1$ (32,220 digits) ranked eighth, while $4648619711505 \times 2^{60000} \pm 1$ (18,075 digits) was nowhere to be seen; $2003663613*2^195000-1$ and $2003663613*2^195000+1$ (58,711 digits), which was discovered on January 15, 2007, by Eric Vautier (from France) of the Twin Prime Search (TPS) project in collaboration with PrimeGrid (BOINC platform), was at the top of the list. As at August 2009, $65516468355 \cdot 2^{333333}-1$ and $65516468355 \cdot 2^{333333}+1$ (100,355 digits) is at the top of the list of top 20 twin primes pairs, while $318032361 \cdot 2^{107001} \pm 1$ (32,220 digits) ranks 11th, and, $2003663613*2^195000-1$ and $2003663613*2^195000+1$ (58,711 digits) ranks second in this list.

We can expect larger twin primes than these extremely large twin primes, much larger ones, infinitely larger ones, to be discovered in due course.

LIST OF PRIMES PAIRS FOR THE FIRST 2,500 CONSECUTIVE PRIMES, 2 TO 22,307, RANKED ACCORDING TO THEIR FREQUENCIES OF APPEARANCE

S. No.	Ranking	Prime Pairs	No. Of Pairs	Percentage
(1)	1	primes pair separated by 6 integers	482	19.29 %
(2)	2	primes pair separated by 4 integers	378	15.13 %
(3)	3	primes pair separated by 2 integers (t. p.)	376	15.05 %
(4)	4	primes pair separated by 12 integers	267	10.68 %
(5)	5	primes pair separated by 10 integers	255	10.20 %
(6)	6	primes pair separated by 8 integers	229	9.16 %
(7)	7	primes pair separated by 14 integers	138	5.52 %
(8)	8	primes pair separated by 18 integers	111	4.44 %
(9)	9	primes pair separated by 16 integers	80	3.20 %
(10)	10	primes pair separated by 20 integers	47	1.88 %
(11)	11	primes pair separated by 22 integers	46	1.84 %
(12)	12	primes pair separated by 30 integers	24	0.96 %
(13)	13	primes pair separated by 28 integers	19	0.76 %
(14)	14	primes pair separated by 24 integers	16	0.64 %
(15)	15	primes pair separated by 26 integers	10	0.40 %
(16)	16	primes pair separated by 34 integers	9	0.36 %
(17)	17	primes pair separated by 36 integers	5	0.20 %
(18)	18	primes pair separated by 32 integers	2	0.08 %
(19)	18	primes pair separated by 40 integers	2	0.08 %
(20)	19	primes pair separated by 42 integers	1	0.04 %
(21)	19	primes pair separated by 52 integers	1	0.04 %

Total No. Of Primes Pairs In List: 2,498

It is evident in the above list that the primes pairs separated by 6 integers, 4 integers and 2 integers (twin primes), among the 21 classifications of primes pairs separated by from 2 integers to 52 integers (primes pairs separated by 38 integers, 44 integers, 46 integers, 48 integers & 50 integers are not among them, but, they are expected to appear further down in the infinite list of the primes), are the most dominant, important. There is a long list of other primes pairs, besides those shown in the above list, which also play a part as the building-blocks of the infinite list of the integers.

The list of the integers is infinite. The list of the primes is also infinite. The infinite primes are the building-blocks of the infinite integers - the infinite odd integers are all either primes or composites of primes, and, the infinite even integers, except for 2 which is a prime, are all also composites of primes. Therefore, all the primes pairs separated by the integers of various magnitudes, as described above, can never all be finite. If there is any possibility at all for any of these primes pairs to be finite, there is only the possibility that a number of these primes pairs are finite (but never all of them). However, will it have to be the primes pairs separated by 2 integers or twin primes (which are the subject of our investigation here), which are the only primes pair, or, one among a number of primes pairs, which are finite? Why question only the infinity of the primes pairs separated by 2 integers, the twin primes? Are not the infinities of the primes pairs separated by 8 integers and more, whose frequencies of appearance are lower, as compared to those of the primes pairs which are separated by 6, 4 and 2 integers respectively, in the above list of primes pairs, more questionable? Why single out only the twin primes? (There are at least 18 other primes pairs, separated by from 8 integers to 52 integers, whose respective infinities should be more suspect, as is evident from the above list of primes pairs, if any infinities should be doubted. Evidently, the primes pairs separated by 2 integers (twin primes) are not that likely to be finite.)

The above represents anecdotal evidence that the twin primes are infinite, which is a ratification of the actual proof given earlier.

APPENDIX 3

The following algorithms will be able to generate or sieve all the twin primes in any range of odd numbers which are all larger than those in the list of known consecutive primes/twin primes; these 2 important algorithms will provide plenty of numerical evidence that the twin primes are infinite:-

Algorithm 1
We would provide an example with Items (1) to (3) from the following list of products of consecutive primes/twin primes, which should be sufficient for our purpose here:-

1) $3 \times 5 = 15$
2) $3 \times 5 \times 7 = 105$
3) $3 \times 5 \times 7 \times 11 = 1,155$
4) $3 \times 5 \times 7 \times 11 \times 13 = 15,015$
5) $3 \times 5 \times 7 \times 11 \times 13 \times 17 = 255,255$
6) $3 \times 5 \times 7 \times 11 \times 13 \times 17 \times 19 = 4,849,845$

.
.

.
.

The example is as follows:-

1) For 3 x 5 = 15, we would find all the consecutive pairs of odd numbers between 5 & 15 which differ from one another by 2 and are not divisible by any of the consecutive primes/twin primes 3 & 5 in the list of consecutive primes/twin primes 3 x 5 whose product is 15.

There is only 1 pair of odd numbers between 5 & 15 which differ from one another by 2 and are not divisible by the consecutive primes/twin primes 3 & 5 in the list of consecutive primes/twin primes 3 x 5 - they are the twin primes 11 & 13.

2) Similarly, for 3 x 5 x 7 = 105, we would find all the consecutive pairs of odd numbers between 7 & 105 which differ from one another by 2 and are not divisible by any of the consecutive primes/twin primes 3, 5 & 7 in the list of consecutive primes/twin primes 3 x 5 x 7 whose product is 105.

The consecutive pairs of odd numbers between 7 & 105 which differ from one another by 2 and are not divisible by the consecutive primes/twin primes 3, 5 & 7 are the following consecutive twin primes:

(a) 11 & 13
(b) 17 & 19
(c) 29 & 31
(d) 41 & 43
(e) 59 & 61
(f) 71 & 73
(g) 101 & 103

3) Similarly, in this final case, for 3 x 5 x 7 x 11 = 1,155, we would find all the consecutive pairs of odd numbers between 11 & 1,155 which differ from one another by 2 and are not divisible by any of the consecutive primes/twin primes 3, 5, 7 & 11 in the list of consecutive primes/twin primes 3 x 5 x 7 x 11 whose product is 1,155.

Many of the consecutive pairs of odd numbers between 11 & 1,155 which differ from one another by 2 and are not divisible by the consecutive primes/twin primes 3, 5, 7 & 11 are twin primes (while the rest are primes larger than 3, 5, 7 & 11 and/or composite numbers whose prime factors are each larger than 3, 5, 7 & 11), some of which are as follows:

(a) 17 & 19
(b) 29 & 31
(c) 41 & 43
(d) 59 & 61
(e) 71 & 73
(f) 101 & 103
(g) 107 & 109
(h) 137 & 139
(i) 149 & 151
(j) 179 & 181
(k) Etc. to 1,151 & 1,153

In this way, we would also be able to achieve the following:-

1) For 3 x 5 x 7 x 11 x 13 = 15,015, find all the consecutive twin primes between 13 and 15,015.
2) For 3 x 5 x 7 x 11 x 13 x 17 = 255,255, find all the consecutive twin primes between 17 and 255,255.
3) For 3 x 5 x 7 x 11 x 13 x 17 x 19 = 4,849,845, find all the consecutive twin primes between 19 and 4,849,845.

.
.
.
.

Algorithm 2

We would, similar to Algorithm 1 above, also provide an example with Items (1) to (3) from the following list of products of consecutive primes/twin primes, which should be sufficient for our purpose here:-

1) 3 x 5 = 15
2) 3 x 5 x 7 = 105
3) 3 x 5 x 7 x 11 = 1,155

4) $3 \times 5 \times 7 \times 11 \times 13 = 15,015$
5) $3 \times 5 \times 7 \times 11 \times 13 \times 17 = 255,255$
6) $3 \times 5 \times 7 \times 11 \times 13 \times 17 \times 19 = 4,849,845$

.
.
.
.
.

The example is as follows:-

1) For $3 \times 5 = 15$, we would first find all the consecutive pairs of even numbers between 5 & 15 which differ from one another by 2 and are not divisible by any of the consecutive primes/twin primes 3 & 5 in the list of consecutive primes/twin primes 3 x 5. Then we deduct each of these consecutive pairs of even numbers which are not divisible by any of the consecutive primes/twin primes 3 & 5 from the product of these consecutive primes/twin primes 3 x 5 which is 15. The results would each be 1 pair of twin primes, 1 prime & 1 composite of primes, or, 2 composites of primes. In this way, we would be able to find all the consecutive twin primes between 5 & 15.

There is only 1 pair of even numbers between 5 & 15 which differ from one another by 2 and are not divisible by any of the consecutive primes/twin primes 3 & 5 in the list of consecutive primes/twin primes 3 x 5 - they are the pair 2 & 4.

The following is the result after we deduct this pair of even numbers 2 & 4 which are not divisible by any of the consecutive primes/twin primes 3 & 5 from the product of these consecutive primes/twin primes 3 x 5 which is 15:

(a) 15 - 2 & 15 - 4: 13 & 11 (twin primes)

2) Similarly, for $3 \times 5 \times 7 = 105$, we would first find all the consecutive pairs of even numbers between 7 & 105 which differ from one another by 2 and are not divisible by any of the consecutive primes/twin primes 3, 5 & 7 in the list of consecutive primes/twin primes 3 x 5 x 7, which are as follows:

(a) 2 & 4
(b) 32 & 34
(c) 44 & 46

(d) 62 & 64
(e) 74 & 76
(f) 86 & 88
(g) 92 & 94

Then we deduct each of these consecutive pairs of even numbers which are not divisible by any of the consecutive primes/twin primes 3, 5 & 7 from the product of these consecutive primes/twin primes 3 x 5 x 7 which is 105. The results would each be 1 pair of twin primes, 1 prime & 1 composite of primes, or, 2 composites of primes. In this way, we would be able to find all the consecutive twin primes between 7 & 105, which are as follows:

(a) 105 - 2 & 105 - 4:   103 & 101 (twin primes)
(b) 105 - 32 & 105 - 34: 73 & 71 (twin primes)
(c) 105 - 44 & 105 - 46: 61 & 59 (twin primes)
(d) 105 - 62 & 105 - 64: 43 & 41 (twin primes)
(e) 105 - 74 & 105 - 76: 31 & 29 (twin primes)
(f) 105 - 86 & 105 - 88: 19 & 17 (twin primes)
(g) 105 - 92 & 105 - 94: 13 & 11 (twin primes)

3) Similarly, in this final case, for 3 x 5 x 7 x 11 = 1,155, we would first find all the consecutive pairs of even numbers between 11 & 1,155 which differ from one another by 2 and are not divisible by any of the consecutive primes/twin primes 3, 5, 7 & 11 in the list of consecutive primes/twin primes 3 x 5 x 7 x 11, some of which are as follows:

(a) 2 & 4
(b) 32 & 34
(c) 62 & 64
(d) 74 & 76
(e) 92 & 94
(f) 116 & 118
(g) 122 & 124
(h) 134 & 136
(i) Etc. to 1,136 & 1,138

Next we deduct each of these consecutive pairs of even numbers which are not divisible by any of the consecutive primes/twin primes 3, 5, 7 & 11 from the product of these consecutive primes/twin primes 3 x 5 x 7 x 11 which is 1,155. The results would each be 1 pair of twin primes, 1 prime & 1 composite of primes, or, 2 composites of primes. In this way, we would be able to find all the consecutive twin primes between 11 & 1,155, some of which are as follows:

(a) 1,155 - 2 & 1,155 - 4:   1,153 & 1,151 (twin primes)
(b) 1,155 - 32 & 1,155 - 34:   1,123 (prime) & 1,121 (composite of primes which
                                                   are each larger than 3, 5, 7 &
                                                   11 = 19 x 59)
(c) 1,155 - 62 & 1,155 - 64:   1,093 & 1,091 (twin primes)
(d) 1,155 - 74 & 1,155 - 76:   1,081   &   1,079
                               (composite of primes   (composite of
                               which are each larger   primes which are
                               than 3, 5, 7 & 11 =   each larger than
                               23 x 47)   3, 5, 7 & 11 =
                                                             13 x 83)
(e) 1,155 - 92 & 1,155 - 94:   1,063 & 1,061 (twin primes)
(f) 1,155 - 116 & 1,155 - 118:   1,039 (prime) & 1,037 (composite of primes which
                                                   are each larger than 3, 5, 7 &
                                                   11 = 17 x 61)
(g) 1,155 - 122 & 1,155 - 124:   1,033 & 1,031 (twin primes)
(h) 1,155 - 134 & 1,155 - 136:   1,021 & 1,019 (twin primes)
(i) Etc. to 1,155 - 1,136 & 1,155 - 1,138: 19 & 17 (twin primes)

In like manner, we would also be able to achieve the following:-

1) For 3 x 5 x 7 x 11 x 13 = 15,015, find all the consecutive twin primes between 13 and 15,015.
2) For 3 x 5 x 7 x 11 x 13 x 17 = 255,255, find all the consecutive twin primes between 17 and 255,255.
3) For 3 x 5 x 7 x 11 x 13 x 17 x 19 = 4,849,845, find all the consecutive twin primes between 19 and 4,849,845.

.

.

.

.

By utilising any of the above algorithms (preferably the evidently more efficient Algorithm 1), we will be able to find many twin primes which are all larger than those in any chosen list of consecutive primes/twin primes, i.e., we will be able to generate many larger and larger twin primes with these algorithms.

It would evidently be difficult to accept a proof of the twin primes conjecture without having to confirm or check the validity of the logic by computing a sufficiently long list of twin primes, even to the extent of looking out for counter-examples. Hence, the great importance of the above algorithms.

APPENDIX 4

Further Remarks On The Twin Primes

We note a very important intrinsic characteristic of the primes. Like all the houses in a neighbourhood or location which are separated from each other by the number of houses between them, the primes are also separated from each other by the number of integers separating them. The closest will of course be the prime neighbours separated by 2 integers (i.e., twin primes), followed next in proximity by the prime neighbours separated by 4 integers, then by the prime neighbours separated by 6 integers, the prime neighbours separated by 8 integers, the prime neighbours separated by 10 integers, the prime neighbours separated by 12 integers, and so on, by larger and larger intervals, as is shown in Appendix 2. The twin primes are actually comparable to 2 closest neighbours living just next door to one another. There will always be 2 closest next-door neighbours, neighbours living 2 doors away, neighbours living 3 doors away, neighbours living 4 doors away, neighbours living 5 doors away, neighbours living 6 doors away, and so on, by greater and greater intervals, in any neighbourhood or residential area; there will always be different intervals separating all the houses in a neighbourhood or location. Similarly, in the infinite list of the primes, there will always be different intervals separating all the primes, ranging from the smallest interval of 2 integers (in the case of the twin primes), 4 integers, 6 integers, 8 integers, 10 integers, 12 integers, and more and more integers, etc., which is an intrinsic characteristic of the primes. In other words, there will always be intervals of various magnitudes or sizes (i.e., intervals of various numbers of integers) between, separating, all the primes in the infinite list of the primes, and, each of these intervals of various magnitudes or sizes can be expected to be infinite as the list of the primes is infinite. The twin primes, which we are examining here, are not likely to be finite (as is evident from Appendix 2), and should be infinite; in fact, to say that the twin primes are finite is like saying that next-door neighbours who are closest are rare and limited, which is absurd.

Hence, the conclusion that the twin primes are infinite.

13 ANOTHER PROOF OF THE TWIN PRIMES CONJECTURE

Many believe that the twin primes are infinite. In fact, twin primes pairs could easily be found among the integers. There is evidently no region of the natural number system so remote that it lies beyond the largest twin primes pair. It is even possible to forecast the approximate number of twin primes pairs found in any region of the natural number system.

The occurrence of twin primes pairs is evidently unpredictable or random. This means that the chance of 2 numbers x and x + 2 being prime (twin primes) is somewhat similar to the chance of getting heads on 2 successive tosses of a coin. If 2 successive tosses of a coin are independent, the chance of success of obtaining heads for the 2 successive tosses of the coin is the product of the chances of success of obtaining a head for each toss of the coin. As each coin has probability ½ of coming up heads with a toss, 2 coins would have probability ½ x ½ = ¼ of coming up a pair of heads with a toss.

The prime number theorem, which had been proven, states that if n is a large number, and we select a number x at random between 0 and n, the chance that x is prime would be approximately 1/log n, the larger n is, the better would be the approximation given by 1/log n to the proportion of primes in the numbers up to n. Like 2 coins coming up heads, the chance that both x and x + 2 are prime (twin primes) would be approximately $1/(\log n)^2$. That is, there would be approximately $n/(\log n)^2$ twin primes pairs between 0 and n. As n goes to infinity, this fraction approaches infinity. This represents a quantitative version of the twin primes conjecture.

As x + 2 being prime depends on the fact that x is already prime, we should modify the estimate $n/(\log n)^2$ to $(1.32032..)n/(\log n)^2$.

The following is a comparison between the twin primes predicted by the above formula and the twin primes found, where the agreement is evidently very good:-

INTERVAL	TWIN PRIMES	
	PREDICTED	FOUND
100,000,000 - 100,150,000	584	601
1,000,000,000 - 1,000,150,000	461	466
10,000,000,000 - 10,000,150,000	374	389
100,000,000,000 - 100,000,150,000	309	276
1,000,000,000,000 - 1,000,000,150,000	259	276
10,000,000,000,000 - 10,000,000,150,000	221	208
100,000,000,000,000 - 100,000,000,150,000	191	186
1,000,000,000,000,000 - 1,000,000,000,150,000	166	161

All this represents numerical evidence that the twin primes are infinite as we could find more twin primes pairs whenever we look for them. But the proof is lacking.

Lemma 1: According to the precepts of fractal geometry and group theory, symmetry is a very important, intrinsic part of nature. There is symmetry all around us and within us. There is evident symmetry in human bodies, the structures of viruses and bacteria, polymers and ceramic materials, the permutations of numbers, the universe and many others, even the movements of prices in financial markets, the growths of populations, the sound of music, the flow of blood through our circulatory system, the behavior of people en masse, etc. In other words, regularity, pattern, order, uniformity or symmetry is evident everywhere.

The reasoning here makes use of a very important idea in fractal geometry and group theory, namely, symmetry.

A prime number is an integer which is divisible only by 1 and itself, e.g., 2, 3, 7, 19, etc. A twin primes pair are 2 primes which differ from one another by 2, e.g., 5 & 7, 11 & 13, 17 & 19, and, 29 & 31, etc. A composite number or non-prime is a product of primes or prime factors, e.g., the composite numbers 15 is the product of 2 primes, 3 and 5 (15 = 3 x 5), and 231 is the product of 3 primes, 3, 7 and 11 (231 = 3 x 7 x 11), etc. The integers or whole numbers are either primes or composites and are infinite.

The primes, which Euclid had proven to be infinite, are the atoms or building-blocks of the infinite integers or whole numbers, which comprise of the infinite list of the odd numbers that are all either primes or products of primes (i.e., composites), and, the infinite list of the even numbers that are all products of primes (i.e., composites, with the exception of 2 which is a prime, e.g., 6 = 2 x 3, 8 = 2 x 2 x 2 and 10 = 2 x 5, etc.). The infinite list of the integers or whole numbers may be classified as an infinite group, with various symmetries, subgroups and infinite elements, hidden within it. These various symmetries, subgroups and infinite elements, within this infinite group may be classified as follows:-

(1) <u>Subgroup A</u>: Infinite consecutive primes such as 2, 3, 5, 7, 11, 13, 17, 19, 23, 29, 31, etc. to infinity ...., separated by 2 integers (twin primes), 4 integers, 6 integers, 8 integers, 10 integers, etc., which, incidentally, except for 2, are all odd numbers; this splitting up of the subgroup into infinite elements is shown below:

   (i) <u>Element A1</u>: Infinite list of all the primes pairs separated by 2 integers (twin primes) .... (Example: 17 & 19)

   (ii) <u>Element A2</u>: Infinite list of all the primes pairs separated by 4 integers/1 odd composite - single composite .... (Example: 79 & 83 separated by 81)

   (iii) <u>Element A3</u>: Infinite list of all the primes pairs separated by 6 integers/2 consecutive odd composites - twin composites .... (Example: 47 & 53 separated by 49 & 51)

(iv) <u>Element A4</u>: Infinite list of all the primes pairs separated by 8 integers/3 consecutive odd composites - "triple" composites ……… . (Example: 359 & 367 separated by 361, 363 & 365)

(v) <u>Element A5</u>: Infinite list of all the primes pairs separated by 10 integers/4 consecutive odd composites - "four-ple" composites ……… . (Example: 709 & 719 separated by 711, 713, 715 & 717)

.
.
.
.

(2) <u>Subgroup B</u>: Infinite consecutive odd composites such as 9, 15, 21, 25, 27, 33, 35, 39, 45, 49, etc. to infinity ……… , of "various sizes" sandwiched between 2 primes; this splitting up of the subgroup into infinite elements is shown below:

(i) <u>Element B1</u>: Infinite list of all "1 odd composite sandwiched between 2 primes - single composite" ……… . (Example: 9 sandwiched between the primes 7 & 11)

(ii) <u>Element B2</u>: Infinite list of all "2 consecutive odd composites sandwiched between 2 primes – twin composites" ……… . (Example: 253 & 255 sandwiched between the primes 251 & 257)

(iii) <u>Element B3</u>: Infinite list of all "3 consecutive odd composites sandwiched between 2 primes – "triple" composites" ……… . (Example: 685, 687 & 689 sandwiched between the primes 683 & 691)

(iv) <u>Element B4</u>: Infinite list of all "4 consecutive odd composites sandwiched between 2 primes – "four-ple" composites" ……… . (Example: 2,769, 2,771, 2,773 & 2,775 sandwiched between the primes 2,767 & 2,777)

(v) <u>Element B5</u>: Infinite list of all "5 consecutive odd composites sandwiched between 2 primes – "five-ple" composites" ……… . (Example: 19,291, 19,293, 19,295, 19,297 & 19,299 sandwiched between the primes 19,289 & 19,301)

.
.
.
.

(3) <u>Subgroup C</u>: Infinite consecutive odd composites separated by 4 integers and 6 integers respectively; this splitting up of the subgroup into the 2 infinite elements is shown below:

(i) Element C1: Infinite list of all "2 consecutive odd composites separated by 4 integers/1 prime" ……….. . (Example: 209 & 213 separated by the prime 211)

(ii) Element C2: Infinite list of all "2 consecutive odd composites separated by 6 integers/2 primes" ……….. . (Example: 279 & 285 separated by the twin primes 281 & 283)

(4) Subgroup D: Infinite single primes and twin primes separating 2 consecutive odd composites; this splitting up of the subgroup into the 2 infinite elements is shown below:

(i) Element D1: Infinite list of all the single primes separating 2 consecutive odd composites ……….. (Example: 23 separating the 2 consecutive odd composites 21 & 25)

(ii) Element D2: Infinite list of all the twin primes separating 2 consecutive odd composites ……….. . (Example: 11 & 13 separating the 2 consecutive odd composites 9 & 15)

(5) Subgroup E: Infinite consecutive even composites such as 4, 6, 8, 10, 12, 14, 16, 18, 20, 22, etc. to infinity ………, all separated by only 2 integers; this subgroup may be classified as a single infinite element ……….. . There is always 1 even number between a twin primes pair, which is separated by 2 integers, a prime and a composite which are separated by 2 integers, and, 2 composites which are separated by 2 integers. That is, the even numbers are always found in Subgroup A, Subgroup B, Subgroup C and Subgroup D above always evenly spaced out in consecutive order by 2 integers.

There is an evident symmetry in the above-mentioned infinite group, which would be broken if any of the elements within were to be finite. There are close interlinks between all the various infinite elements in all the five subgroups above, e.g., the infinity of the list of all the primes pairs each separated by 6 integers (Element A3/Subgroup A) implies the infinity of the list of all the "2 consecutive odd composites sandwiched between 2 primes - twin composites" (Element B2/Subgroup B) and vice versa, the infinity of the list of all the primes pairs each separated by 2 integers (twin primes) (Element A1/SubgroupA) implies the infinity of the list of all the "2 consecutive odd composites separated by 6 integers/2 primes" (Element C2/Subgroup C) and vice versa, the infinity of the list of all the infinite elements (A1, A2, A3, etc. to infinity) in Subgroup A above, which represents the infinity of the list of the primes which Euclid had in fact proven, implies the infinity of the list of all the "2 consecutive odd composites separated by 4 integers/1 prime" (Element C1/Subgroup C) and vice versa, the infinity of the list of all the "2 consecutive odd composites separated by 6 integers/2 primes" (Element C2/Subgroup C) implies the infinity of the list of all the twin primes separating 2 consecutive odd composites (Element D2/Subgroup D) and vice versa, the infinity of all the lists of all the infinite elements (A1, A2, A3 ……, B1, B2, B3 ……, C1 & C2, D1 &

D2) in Subgroup A, Subgroup B, Subgroup C and Subgroup D above implies the infinity of the list of the consecutive even composites, i.e., 4, 6, 8, 10, 12, 14, 16, 18, 20, 22, etc. to infinity (Subgroup E), which we know to be true in any case, and vice versa, etc.

Subgroup A and Subgroup B above are practically "mirror" images of one another - they represent the viewing of the primes and the composites from 2 variant angles - the infinitude, or, finiteness of either implies the infinitude, or, finiteness of the other; the same applies to both Subgroup C and Subgroup D above. It is similar to the following way of viewing a glass which is partially filled: this glass could be described as "half full" or "half empty" if it is half filled, "three-quarter full" or "one-quarter empty" if it is three-quarter filled, or, "one-quarter full" or "three-quarter empty" if it is one-quarter filled, etc.
It is evident that the infinitude, or, finiteness of any one of the above-mentioned elements would imply the infinitude, or, finiteness of the other element that is interlinked with it and vice versa. All these infinite elements are evidently entangled together and complementary, being all the infinite building-blocks of the infinite integers or whole numbers. The infinity of the list of the integers or whole numbers, the primes included, in fact implies that all these various elements within it are infinite, and, vice versa, since all these various elements are closely interlinked and could not do without each other. Therefore, the breaking of the evident intrinsic symmetry of this whole infinite group, i.e., the infinite list of the integers or whole numbers, due to the finiteness of any of the elements within it, could not be possible.

We pose a very important question: Besides questioning whether the infinite list of all the primes pairs separated by 2 integers (twin primes) is really infinite, should we not also be questioning whether the following are really infinite?:

(a) Infinite lists of all the primes pairs separated respectively by 4 integers, 6 integers, 8 integers, 10 integers and sequentially larger integers (as in Subgroup A above).
(b) Infinite lists of all the respective consecutive odd composites of "various sizes" sandwiched between 2 primes (as in Subgroup B above).
(c) The 2 infinite lists with respectively "2 consecutive odd composites separated by 4 integers/1 prime" and "2 consecutive odd composites separated by 6 integers/2 primes" (as in Subgroup C above).
(d) The 2 infinite lists of respective single primes and twin primes separating 2 consecutive odd composites (as in Subgroup D above).
(e) Infinite list of the consecutive even composites all separated by only 2 integers (as in Subgroup E above).

Could there possibly be any symmetry-breaking in the above-mentioned infinite group whence one or more of the elements within it would be finite? In particular, could there be a possibility for the symmetry of this infinite group to be broken due to the finiteness of Element A1 (i.e., the finiteness of the twin primes) within it? Since the above-mentioned group, i.e., the list of the integers or whole numbers, is infinite, it is indeed not possible for all of these elements to be finite. And, there is no evident reason to account for why any of these elements, especially Element A1, i.e., the list of primes separated by 2 integers, or, twin primes, should be finite. In fact, all these infinite elements are like the slabs of various sizes in a building. They are all necessary for the construction of the infinite building known as the "infinite list of the integers or whole numbers" and should thus all be infinite, wherein the symmetry of the infinite group, i.e., the infinite list of the integers or whole numbers, would be preserved.

Therefore, by Lemma 1, all the elements in Subgroup A, Subgroup B, Subgroup C, Subgroup D and Subgroup E above are infinite.

Lemma 2: The Fundamental Theorem of Arithmetic or Unique Factorisation Theorem states that there is only one possible combination of primes which will multiply together to produce any particular composite number, e.g., the only combination of primes which will produce the composite number 2,079 is: 3 x 3 x 3 x 7 x 11. In the same manner, the following composite numbers are also uniquely factorised:

(1) 63 = 3 x 3 x 7 (only)
(2) 153 = 3 x 3 x 17 (only)
(3) 1,021,020 = 2 x 2 x 3 x 5 x 7 x 11 x 13 x 17 (only)

In other words, every positive whole number which is not prime (i.e., every positive whole number which is composite) can be broken up into prime factors, and, this can happen in only 1 way:

$$c \quad = \quad \prod_{p \text{ prime}} p \quad \text{(in only 1 way)}$$

The 10 consecutive twin primes 3 & 5 to 107 & 109, e.g., give rise to the following 10 composite numbers which can be factorised in only 1 way, i.e., can be factorised only by the respective twin primes:

(1) 15 = 3 x 5 (only)
(2) 35 = 5 x 7 (only)
(3) 143 = 11 x 13 (only)
(4) 323 = 17 x 19 (only)

(5) 899 = 29 x 31 (only)
(6) 1,763 = 41 x 43 (only)
(7) 3,599 = 59 x 61 (only)
(8) 5,183 = 71 x 73 (only)
(9) 10,403 = 101 x 103 (only)
(10) 11,663 = 107 x 109 (only)

.

.

.

.

As the composite numbers are infinite, this implies that there is an infinitude of twin primes acting as prime factors for an infinitude of composite numbers in only 1 way as the twin primes are indispensable, i.e., necessary, as prime factors for the formation of an infinite number of composite numbers which can only be formed through the product of twin primes in only 1 way - the twin primes can never be substituted as prime factors of these composite numbers by other primes.

14 A PROOF OF THE GOLDBACH CONJECTURE

The expected mode of solving the Goldbach conjecture appears to be the utilization of advanced calculus or analysis, e.g., by the summation, or, integration, of the reciprocals involving directly or indirectly the primes to see whether they converge or diverge, in order to get a "feel" of the pattern of the distribution of the primes. But, such a method of solving the problem has evidently not succeeded so far. Some other approach or approaches could be more appropriate. This chapter addresses the problem from several different angles, with reasoning backed by quantities that can be checked.

Every even number after 2 is the sum of 2 odd numbers. Every odd number is either a prime which is odd or a composite - product of primes which are odd; notably, every prime with the exception of 2 is an odd number. Every even number after 2 is also a composite, but, a composite with at least 1 even prime factor, namely, 2, while the rest of its prime factors are odd, i.e., it is an even composite.

Therefore, every even number after 2 is the sum of 2 primes which are odd and/or the sum of 1 prime which is odd and 1 odd composite whose prime factors are odd and/or the sum of 2 odd composites whose prime factors are odd, besides being an even composite with at least 1 even prime factor, namely, 2, while the rest of its prime factors are odd.

Lemma:
By Euclid's proof, the primes are infinite; this implies that there would be an infinitude of sums of 2 primes as per the Goldbach conjecture. The even numbers, which are sums of 2 primes as per the conjecture, are also infinite. Thus, there are an infinite number of even numbers which are sums of 2 primes, both the even numbers and sums of 2 primes being infinite.

Corollary:
The odd numbers, which are either prime, every prime with the exception of 2 being an odd number, or composite (have prime factors which are odd), are infinite; this implies that there would be an infinite number of sums of 2 odd numbers, each of which is equal to an even number. Hence, as there is an infinitude of even numbers which are sums of 2 primes, as per the above lemma, and as all primes with the exception of 2 are odd numbers, there are an infinite number of even numbers which are sums of 2 odd numbers that are prime, all the even numbers, sums of 2 odd numbers and primes being infinite; i.e., every even number after 2 is also the sum of 2 odd numbers that are prime.

We thereby see the close interlink or relationship between the primes, even numbers and odd numbers, which are all infinite, which is significant.

The following are thus evident:

a) Every sum of 2 primes which are odd numbers is equal to an even number, as is below in consecutive order:

$2 + 2 = 1 + 3 = \mathbf{4}$
$3 + 3 = 1 + 5 = \mathbf{6}$
$3 + 5 = 1 + 7 = \mathbf{8}$
$5 + 5 = 3 + 7 = \mathbf{10}$
$5 + 7 = 1 + 11 = \mathbf{12}$
$7 + 7 = 3 + 11 = 1 + 13 = \mathbf{14}$
$3 + 13 = 5 + 11 = \mathbf{16}$
$7 + 11 = 5 + 13 = 1 + 17 = \mathbf{18}$
$7 + 13 = 3 + 17 = 1 + 19 = \mathbf{20}$
$11 + 11 = 3 + 19 = 5 + 17 = 11 + 11 = \mathbf{22}$
$11 + 13 = 5 + 19 = 7 + 17 = 1 + 23 = \mathbf{24}$
$13 + 13 = 3 + 23 = 7 + 19 = \mathbf{26}$
$11 + 17 = 5 + 23 = \mathbf{28}$
$13 + 17 = 11 + 19 = 7 + 23 = 1 + 29 = \mathbf{30}$
$3 + 29 = 13 + 19 = 1 + 31 = \mathbf{32}$
$17 + 17 = 3 + 31 = 5 + 29 = 11 + 23 = 17 + 17 = \mathbf{34}$
$17 + 19 = 5 + 31 = 7 + 29 = 13 + 23 = \mathbf{36}$
$19 + 19 = 7 + 31 = 1 + 37 = \mathbf{38}$
$3 + 37 = 11 + 29 = 17 + 23 = \mathbf{40}$
$19 + 23 = 5 + 37 = 11 + 31 = 13 + 29 = 1 + 41 = \mathbf{42}$
$3 + 41 = 7 + 37 = 13 + 31 = 1 + 43 = \mathbf{44}$
$23 + 23 = 3 + 43 = 5 + 41 = 17 + 29 = \mathbf{46}$
$5 + 43 = 7 + 41 = 11 + 37 = 17 + 31 = 19 + 29 = 1 + 47 = \mathbf{48}$
$3 + 47 = 7 + 43 = 13 + 37 = 19 + 31 = \mathbf{50}$
$23 + 29 = 5 + 47 = 11 + 41 = \mathbf{52}$
$7 + 47 = 11 + 43 = 13 + 41 = 17 + 37 = 23 + 31 = 1 + 53 = \mathbf{54}$
$3 + 53 = 13 + 43 = 19 + 37 = \mathbf{56}$
$29 + 29 = 5 + 53 = 11 + 47 = 17 + 41 = 29 + 29 = \mathbf{58}$
$29 + 31 = 7 + 53 = 13 + 47 = 17 + 43 = 19 + 41 = 23 + 37 = 1 + 59 = \mathbf{60}$
$31 + 31 = 3 + 59 = 19 + 43 = 1 + 61 = \mathbf{62}$
$3 + 61 = 5 + 59 = 11 + 53 = 17 + 47 = 23 + 41 = \mathbf{64}$
$5 + 61 = 7 + 59 = 13 + 53 = 19 + 47 = 23 + 43 = 29 + 37 = \mathbf{66}$

$7 + 61 = 31 + 37 = 1 + 67 = \mathbf{68}$

$3 + 67 = 11 + 59 = 17 + 53 = 23 + 47 = 29 + 41 = \mathbf{70}$

$5 + 67 = 11 + 61 = 13 + 59 = 19 + 53 = 29 + 43 = 31 + 41 = 1 + 71 = \mathbf{72}$

$37 + 37 = 3 + 71 = 7 + 67 = 13 + 61 = 31 + 43 = 37 + 37 = 1 + 73 = \mathbf{74}$

$3 + 73 = 5 + 71 = 17 + 59 = 23 + 53 = 29 + 47 = \mathbf{76}$

$37 + 41 = 5 + 73 = 7 + 71 = 11 + 67 = 31 + 47 = 37 + 41 = \mathbf{78}$

$7 + 73 = 13 + 67 = 19 + 61 = 37 + 43 = 1 + 79 = \mathbf{80}$

$41 + 41 = 3 + 79 = 11 + 71 = 23 + 59 = 29 + 53 = \mathbf{82}$

$41 + 43 = 5 + 79 = 11 + 73 = 13 + 71 = 17 + 67 = 23 + 61 = 31 + 53 = 37 + 47 = 1 + 83 = \mathbf{84}$

$43 + 43 = 3 + 83 = 7 + 79 = 13 + 73 = 19 + 67 = 43 + 43 = \mathbf{86}$

$5 + 83 = 17 + 71 = 29 + 59 = 41 + 47 = \mathbf{88}$

$7 + 83 = 11 + 79 = 17 + 73 = 19 + 71 = 23 + 67 = 29 + 61 = 31 + 59 = 37 + 53 = 43 + 47 = 1 + 89 = \mathbf{90}$

$3 + 89 = 13 + 79 = 19 + 73 = 31 + 61 = 1 + 91 = \mathbf{92}$

$47 + 47 = 5 + 89 = 11 + 83 = 23 + 71 = 41 + 53 = 47 + 47 = \mathbf{94}$

$5 + 91 = 7 + 89 = 13 + 83 = 17 + 79 = 23 + 73 = 29 + 67 = 37 + 59 = 43 + 53 = \mathbf{96}$

$7 + 91 = 19 + 79 = 31 + 67 = 37 + 61 = 1 + 97 = \mathbf{98}$

$47 + 53 = 3 + 97 = 11 + 89 = 17 + 83 = 29 + 71 = 41 + 59 = 47 + 53 = \mathbf{100}$

$5 + 97 = 11 + 91 = 13 + 89 = 19 + 83 = 23 + 79 = 29 + 73 = 31 + 71 = 41 + 61 = 43 + 59 = 1 + 101 = \mathbf{102}$

.
.
.
.

b) Every sum of 1 prime which is an odd number & 1 odd composite which is the product of primes which are odd, is equal to the sum of 2 primes which are odd numbers, which are all each equal to an even number, as is below in consecutive order:

$\mathbf{3} + \mathbf{9} = 5 + 7 = 1 + 11 = \mathbf{12}$

$\mathbf{5} + \mathbf{9} = 3 + 11 = 7 + 7 = 1 + 13 = \mathbf{14}$

$7 + 9 = 3 + 13 = 5 + 11 = \mathbf{16}$

$\mathbf{3} + \mathbf{15} = 7 + 11 = 5 + 13 = 1 + 17 = \mathbf{18}$

$\mathbf{11} + \mathbf{9} = 3 + 17 = 7 + 13 = 1 + 19 = \mathbf{20}$

$\mathbf{13} + \mathbf{9} = 3 + 19 = 5 + 17 = 11 + 11 = \mathbf{22}$

$\mathbf{3} + \mathbf{21} = 11 + 13 = 5 + 19 = 7 + 17 = 1 + 23 = \mathbf{24}$

$\mathbf{17} + \mathbf{9} = 3 + 23 = 7 + 19 = 13 + 13 = \mathbf{26}$

19 + **9** = 5 + 23 = 11 + 17 = **28**
5 + **25** = 13 + 17 = 11 + 19 = 7 + 23 = 1 + 29 = **30**
23 + **9** = 3 + 29 = 13 + 19 = 1 + 31 = **32**
7 + **27** = 17 + 17 = 3 + 31 = 5 + 29 = 11 + 23 = 17 + 17 = **34**
3 + **33** = 17 + 19 = 5 + 31 = 7 + 29 = 13 + 23 = **36**
29 + **9** = 7 + 31 = 19 + 19 = 1 + 37 = **38**
31 + **9** = 3 + 37 = 11 + 29 = 17 + 23 = **40**
3 + **39** = 19 + 23 = 5 + 37 = 11 + 31 = 13 + 29 = 1 + 41 = **42**
5 + **39** = 3 + 41 = 7 + 37 = 13 + 31 = 1 + 43 = **44**
37 + **9** = 3 + 43 = 5 + 41 = 17 + 29 = 23 + 23 = **46**
3 + **45** = 5 + 43 = 7 + 41 = 11 + 37 = 17 + 31 = 19 + 29 = 1 + 47 = **48**
41 + **9** = 3 + 47 = 7 + 43 = 13 + 37 = 19 + 31 = **50**
43 + **9** = 5 + 47 = 11 + 41 = 23 + 29 = **52**
5 + **49** = 7 + 47 = 11 + 43 = 13 + 41 = 17 + 37 = 23 + 31 = 1 + 53 = **54**
47 + **9** = 3 + 53 = 13 + 43 = 19 + 37 = **56**
3 + **55** = 29 + 29 = 5 + 53 = 11 + 47 = 17 + 41 = 29 + 29 = **58**
5 + **55** = 29 + 31 = 7 + 53 = 13 + 47 = 17 + 43 = 19 + 41 = 23 + 37 = 1 + 59 = **60**
53 + **9** = 3 + 59 = 19 + 43 = 31 + 31 = 1 + 61 = **62**
7 + **57** = 3 + 61 = 5 + 59 = 11 + 53 = 17 + 47 = 23 + 41 = **64**
11 + **55** = 5 + 61 = 7 + 59 = 13 + 53 = 19 + 47 = 23 + 43 = 29 + 37 = **66**
59 + **9** = 7 + 61 = 31 + 37 = 1 + 67 = **68**
61 + **9** = 3 + 67 = 11 + 59 = 17 + 53 = 23 + 47 = 29 + 41 = **70**
3 + **69** = 5 + 67 = 11 + 61 = 13 + 59 = 19 + 53 = 29 + 43 = 31 + 41 = 1 + 71 = **72**
5 + **69** = 37 + 37 = 3 + 71 = 7 + 67 = 13 + 61 = 31 + 43 = 37 + 37 = 1 + 73 = **74**
67 + **9** = 3 + 73 = 5 + 71 = 17 + 59 = 23 + 53 = 29 + 47 = **76**
3 + **75** = 37 + 41 = 5 + 73 = 7 + 71 = 11 + 67 = 31 + 47 = 37 + 41 = **78**
71 + **9** = 7 + 73 = 13 + 67 = 19 + 61 = 37 + 43 = 1 + 79 = **80**
73 + **9** = 3 + 79 = 11 + 71 = 23 + 59 = 29 + 53 = 41 + 41 = **82**
3 + **81** = 41 + 43 = 5 + 79 = 11 + 73 = 13 + 71 = 17 + 67 = 23 + 61 = 31 + 53 = 37 + 47 = 1+ 83 = **84**
5 + **81** = 43 + 43 = 3 + 83 = 7 + 79 = 13 + 73 = 19 + 67 = 43 + 43 = **86**
79 + **9** = 5 + 83 = 17 + 71 = 29 + 59 = 41 + 47 = **88**
3 + **87** = 7 + 83 = 11 + 79 = 17 + 73 = 19 + 71 = 23 + 67 = 29 + 61 = 31 + 59 = 37 + 53 = 43 + 47 = 1 + 89 = **90**
83 + **9** = 3 + 89 = 13 + 79 = 19 + 73 = 31 + 61 = 1 + 91 = **92**
7 + **87** = 47 + 47 = 5 + 89 = 11 + 83 = 23 + 71 = 41 + 53 = 47 + 47 = **94**
3 + **93** = 5 + 91 = 7 + 89 = 13 + 83 = 17 + 79 = 23 + 73 = 29 + 67 = 37 + 59 = 43 + 53 = **96**
89 + **9** = 7 + 91 = 19 + 79 = 31 + 67 = 37 + 61 = 1 + 97 = **98**

$91 + 9 = 3 + 97 = 11 + 89 = 17 + 83 = 29 + 71 = 41 + 59 = 47 + 53 = \mathbf{100}$

$3 + 99 = 5 + 97 = 11 + 91 = 13 + 89 = 19 + 83 = 23 + 79 = 29 + 73 = 31 + 71 = 41 + 61 = 43 + 59 = 1 + 101 = \mathbf{102}$

.

.

.

.

c) Every sum of 2 odd composites which are products of primes which are odd, is equal to the sum of 2 primes which are odd numbers, which are all each equal to an even number, as is below in consecutive order:

$\mathbf{9} + \mathbf{9} = 5 + 13 = 7 + 11 = 1 + 17 = \mathbf{18}$

$\mathbf{9} + \mathbf{15} = 5 + 19 = 7 + 17 = 11 + 13 = 1 + 23 = \mathbf{24}$

$\mathbf{15} + \mathbf{15} = 7 + 23 = 11 + 19 = 13 + 17 = 1 + 29 = \mathbf{30}$

$\mathbf{9} + \mathbf{25} = 7 + 27 = 17 + 17 = 3 + 31 = 5 + 29 = 11 + 23 = 17 + 17 = \mathbf{34}$

$\mathbf{15} + \mathbf{21} = 5 + 31 = 7 + 29 = 13 + 23 = 17 + 19 = \mathbf{36}$

$\mathbf{15} + \mathbf{25} = 3 + 37 = 11 + 29 = 17 + 23 = \mathbf{40}$

$\mathbf{21} + \mathbf{21} = 5 + 37 = 11 + 31 = 13 + 29 = 19 + 23 = 1 + 41 = \mathbf{42}$

$\mathbf{9} + \mathbf{35} = 3 + 41 = 7 + 37 = 13 + 31 = 1 + 43 = \mathbf{44}$

$\mathbf{21} + \mathbf{25} = 3 + 43 = 5 + 41 = 17 + 29 = 23 + 23 = \mathbf{46}$

$\mathbf{9} + \mathbf{39} = 5 + 43 = 7 + 41 = 11 + 37 = 17 + 31 = 19 + 29 = 1 + 47 = \mathbf{48}$

$\mathbf{25} + \mathbf{25} = 3 + 47 = 7 + 43 = 13 + 37 = 19 + 31 = \mathbf{50}$

$\mathbf{25} + \mathbf{27} = 5 + 47 = 11 + 41 = 23 + 29 = \mathbf{52}$

$\mathbf{27} + \mathbf{27} = 7 + 47 = 11 + 43 = 13 + 41 = 17 + 37 = 23 + 31 = 1 + 53 = \mathbf{54}$

$\mathbf{21} + \mathbf{35} = 3 + 53 = 13 + 43 = 19 + 37 = \mathbf{56}$

$\mathbf{9} + \mathbf{49} = 29 + 29 = 5 + 53 = 11 + 47 = 17 + 41 = 29 + 29 = \mathbf{58}$

$\mathbf{27} + \mathbf{33} = 7 + 53 = 13 + 47 = 17 + 43 = 19 + 41 = 23 + 37 = 29 + 31 = 1 + 59 = \mathbf{60}$

$\mathbf{27} + \mathbf{35} = 31 + 31 = 3 + 59 = 19 + 43 = 1 + 61 = \mathbf{62}$

$\mathbf{9} + \mathbf{55} = 3 + 61 = 5 + 59 = 11 + 53 = 17 + 47 = 23 + 41 = \mathbf{64}$

$\mathbf{33} + \mathbf{33} = 5 + 61 = 7 + 59 = 13 + 53 = 19 + 47 = 23 + 43 = 29 + 37 = \mathbf{66}$

$\mathbf{33} + \mathbf{35} = 7 + 61 = 31 + 37 = 1 + 67 = \mathbf{68}$

$\mathbf{35} + \mathbf{35} = 3 + 67 = 11 + 59 = 17 + 53 = 23 + 47 = 29 + 41 = \mathbf{70}$

$\mathbf{9} + \mathbf{63} = 5 + 67 = 11 + 61 = 13 + 59 = 19 + 53 = 29 + 43 = 31 + 41 = 1 + 71 = \mathbf{72}$

$\mathbf{35} + \mathbf{39} = 3 + 71 = 7 + 67 = 13 + 61 = 31 + 43 = 37 + 37 = 1 + 73 = \mathbf{74}$

$\mathbf{21} + \mathbf{55} = 3 + 73 = 5 + 71 = 17 + 59 = 23 + 53 = 29 + 47 = \mathbf{76}$

$\mathbf{39} + \mathbf{39} = 5 + 73 = 7 + 71 = 11 + 67 = 31 + 47 = 37 + 41 = \mathbf{78}$

$15 + 65 = 7 + 73 = 13 + 67 = 19 + 61 = 37 + 43 = 1 + 79 = \mathbf{80}$

$25 + 57 = 41 + 41 = 3 + 79 = 11 + 71 = 23 + 59 = 29 + 53 = \mathbf{82}$

$39 + 45 = 5 + 79 = 11 + 73 = 13 + 71 = 17 + 67 = 23 + 61 = 31 + 53 = 37 + 47 = 41 + 43 = 1 + 83 = \mathbf{84}$

$9 + 77 = 43 + 43 = 3 + 83 = 7 + 79 = 13 + 73 = 19 + 67 = 43 + 43 = \mathbf{86}$

$25 + 63 = 5 + 83 = 17 + 71 = 29 + 59 = 41 + 47 = \mathbf{88}$

$45 + 45 = 7 + 83 = 11 + 79 = 17 + 73 = 19 + 71 = 23 + 67 = 29 + 61 = 31 + 59 = 37 + 53 = 43 + 47 = 1 + 89 = \mathbf{90}$

$15 + 77 = 3 + 89 = 13 + 79 = 19 + 73 = 31 + 61 = 1 + 91 = \mathbf{92}$

$45 + 49 = 5 + 89 = 11 + 83 = 23 + 71 = 41 + 53 = 47 + 47 = \mathbf{94}$

$9 + 87 = 5 + 91 = 7 + 89 = 13 + 83 = 17 + 79 = 23 + 73 = 29 + 67 = 37 + 59 = 43 + 53 = \mathbf{96}$

$49 + 49 = 7 + 91 = 19 + 79 = 31 + 67 = 37 + 61 = 1 + 97 = \mathbf{98}$

$49 + 51 = 3 + 97 = 11 + 89 = 17 + 83 = 29 + 71 = 41 + 59 = 47 + 53 = \mathbf{100}$

$51 + 51 = 5 + 97 = 11 + 91 = 13 + 89 = 19 + 83 = 23 + 79 = 29 + 73 = 31 + 71 = 41 + 61 = 43 + 59 = 1 + 101 = \mathbf{102}$

.
.
.
.

d) From (a), (b) & (c) above, we have the even numbers from 4 to 102 … composed as follows:

1) $\mathbf{4} = 2 + 2 = 1 + 3$ (sum of 2 primes only)
2) $\mathbf{6} = 3 + 3 = 1 + 5$ (sum of 2 primes only)
3) $\mathbf{8} = 3 + 5 = 1 + 7$ (sum of 2 primes only)
4) $\mathbf{10} = 5 + 5 = 3 + 7$ (sum of 2 primes only)
5) $\mathbf{12} = 5 + 7 = 1 + 11 = \mathbf{3 + 9}$ (sum of 1 prime & 1 odd composite)
6) $\mathbf{14} = 3 + 11 = 7 + 7 = 1 + 13 = \mathbf{5 + 9}$ (sum of 1 prime & 1 odd composite)
7) $\mathbf{16} = 3 + 13 = 5 + 11 = \mathbf{7 + 9}$ (sum of 1 prime & 1 odd composite)
8) $\mathbf{18} = 5 + 13 = 7 + 11 = 1 + 17 = \mathbf{3 + 15}$ (sum of 1 prime & 1 odd composite) = $\mathbf{9 + 9}$ (sum of 2 odd composites)
9) $\mathbf{20} = 3 + 17 = 7 + 13 = 1 + 19 = \mathbf{11 + 9}$ (sum of 1 prime & 1 odd composite)
10) $\mathbf{22} = 3 + 19 = 5 + 17 = 11 + 11 = \mathbf{13 + 9}$ (sum of 1 prime & 1 odd composite)
11) $\mathbf{24} = 5 + 19 = 7 + 17 = 11 + 13 = 1 + 23 = \mathbf{3 + 21}$ (sum of 1 prime & 1 odd composite) = $\mathbf{9 + 15}$ (sum of 2 odd composites)
12) $\mathbf{26} = 3 + 23 = 7 + 19 = 13 + 13 = \mathbf{17 + 9}$ (sum of 1 prime & 1 odd composite)
13) $\mathbf{28} = 5 + 23 = 11 + 17 = \mathbf{19 + 9}$ (sum of 1 prime & 1 odd composite)

14) $30 = 7 + 23 = 11 + 19 = 13 + 17 = 1 + 29 = \mathbf{5} + \mathbf{25}$ (sum of 1 prime & 1 odd composite) $= \mathbf{15} + \mathbf{15}$ (sum of 2 odd composites)

15) $32 = 3 + 29 = 13 + 19 = 1 + 31 = \mathbf{23} + \mathbf{9}$ (sum of 1 prime & 1 odd composite)

16) $34 = 17 + 17 = 3 + 31 = 5 + 29 = 11 + 23 = 17 + 17 = \mathbf{7} + \mathbf{27}$ (sum of 1 prime & 1 odd composite) $= \mathbf{9} + \mathbf{25}$ (sum of 2 odd composites)

17) $36 = 5 + 31 = 7 + 29 = 13 + 23 = 17 + 19 = \mathbf{3} + \mathbf{33}$ (sum of 1 prime & 1 odd composite) $= \mathbf{15} + \mathbf{21}$ (sum of 2 odd composites)

18) $38 = 7 + 31 = 19 + 19 = 1 + 37 = \mathbf{29} + \mathbf{9}$ (sum of 1 prime & 1 odd composite)

19) $40 = 3 + 37 = 11 + 29 = 17 + 23 = \mathbf{31} + \mathbf{9}$ (sum of 1 prime & 1 odd composite) $= \mathbf{15} + \mathbf{25}$ (sum of 2 odd composites)

20) $42 = 5 + 37 = 11 + 31 = 13 + 29 = 19 + 23 = 1 + 41 = \mathbf{3} + \mathbf{39}$ (sum of 1 prime & 1 odd composite) $= \mathbf{21} + \mathbf{21}$ (sum of 2 odd composites)

21) $44 = 3 + 41 = 7 + 37 = 13 + 31 = 1 + 43 = \mathbf{5} + \mathbf{39}$ (sum of 1 prime & 1 odd composite) $= \mathbf{9} + \mathbf{35}$ (sum of 2 odd composites)

22) $46 = 3 + 43 = 5 + 41 = 17 + 29 = 23 + 23 = \mathbf{37} + \mathbf{9}$ (sum of 1 prime & 1 odd composite) $= \mathbf{21} + \mathbf{25}$ (sum of 2 odd composites)

23) $48 = 5 + 43 = 7 + 41 = 11 + 37 = 17 + 31 = 19 + 29 = 1 + 47 = \mathbf{3} + \mathbf{45}$ (sum of 1 prime & 1 odd composite) $= \mathbf{9} + \mathbf{39}$ (sum of 2 odd composites)

24) $50 = 3 + 47 = 7 + 43 = 13 + 37 = 19 + 31 = \mathbf{41} + \mathbf{9}$ (sum of 1 prime & 1 odd composite) $= \mathbf{25} + \mathbf{25}$ (sum of 2 odd composites)

25) $52 = 5 + 47 = 11 + 41 = 23 + 29 = = \mathbf{43} + \mathbf{9}$ (sum of 1 prime & 1 odd composite) $= \mathbf{25} + \mathbf{27}$ (sum of 2 odd composites)

26) $54 = 7 + 47 = 11 + 43 = 13 + 41 = 17 + 37 = 23 + 31 = 1 + 53 = \mathbf{5} + \mathbf{49}$ (sum of 1 prime & 1 odd composite) $= \mathbf{27} + \mathbf{27}$ (sum of 2 odd composites)

27) $56 = 3 + 53 = 13 + 43 = 19 + 37 = \mathbf{47} + \mathbf{9}$ (sum of 1 prime & 1 odd composite) $= \mathbf{21} + \mathbf{35}$ (sum of 2 odd composites)

28) $58 = 29 + 29 = 5 + 53 = 11 + 47 = 17 + 41 = 29 + 29 = \mathbf{3} + \mathbf{55}$ (sum of 1 prime & 1 odd composite) $= \mathbf{9} + \mathbf{49}$ (sum of 2 odd composites)

29) $60 = 7 + 53 = 13 + 47 = 17 + 43 = 19 + 41 = 23 + 37 = 29 + 31 = 1 + 59 = \mathbf{5} + \mathbf{55}$ (sum of 1 prime & 1 odd composite) $= \mathbf{27} + \mathbf{33}$ (sum of 2 odd composites)

30) $62 = 3 + 59 = 19 + 43 = 31 + 31 = 1 + 61 = \mathbf{53} + \mathbf{9}$ (sum of 1 prime & 1 odd composite) $= \mathbf{27} + \mathbf{35}$ (sum of 2 odd composites)

31) $64 = 3 + 61 = 5 + 59 = 11 + 53 = 17 + 47 = 23 + 41 = \mathbf{7} + \mathbf{57}$ (sum of 1 prime & 1 odd composite) $= \mathbf{9} + \mathbf{55}$ (sum of 2 odd composites)

32) $66 = 5 + 61 = 7 + 59 = 13 + 53 = 19 + 47 = 23 + 43 = 29 + 37 = \mathbf{11} + \mathbf{55}$ (sum of 1 prime & 1 odd composite) $= \mathbf{33} + \mathbf{33}$ (sum of 2 odd composites)

33) $68 = 7 + 61 = 31 + 37 = 1 + 67 = \mathbf{59} + \mathbf{9}$ (sum of 1 prime & 1 odd composite) $= \mathbf{33} + \mathbf{35}$ (sum of 2

odd composites)

34) **70** = 3 + 67 = 11 + 59 = 17 + 53 = 23 + 47 = 29 + 41 = **61 + 9** (sum of 1 prime & 1 odd composite) = **35 + 35** (sum of 2 odd composites)

35) **72** = 5 + 67 = 11 + 61 = 13 + 59 = 19 + 53 = 29 + 43 = 31 + 41 = 1 + 71 = **3 + 69** (sum of 1 prime & 1 odd composite) = **9 + 63** (sum of 2 odd composites)

36) **74** = 3 + 71 = 7 + 67 = 13 + 61 = 31 + 43 = 37 + 37 = 1 + 73 = **5 + 69** (sum of 1 prime & 1 odd composite) = **35 + 39** (sum of 2 odd composites)

37) **76** = 3 + 73 = 5 + 71 = 17 + 59 = 23 + 53 = 29 + 47 = **67 + 9** (sum of 1 prime & 1 odd composite) = **21 + 55** (sum of 2 odd composites)

38) **78** = 5 + 73 = 7 + 71 = 11 + 67 = 31 + 47 = 37 + 41 = **3 + 75** (sum of 1 prime & 1 odd composite) = **39 + 39** (sum of 2 odd composites)

39) **80** = 7 + 73 = 13 + 67 = 19 + 61 = 37 + 43 = 1 + 79 = **71 + 9** (sum of 1 prime & 1 odd composite) = **15 + 65** (sum of 2 odd composites)

40) **82** = 3 + 79 = 11 + 71 = 23 + 59 = 29 + 53 = 41 + 41 = **73 + 9** (sum of 1 prime & 1 odd composite) = **25 + 57** (sum of 2 odd composites)

41) **84** = 5 + 79 = 11 + 73 = 13 + 71 = 17 + 67 = 23 + 61 = 31 + 53 = 37 + 47 = 41 + 43 = 1 + 83 = **3 + 81** (sum of 1 prime & 1 odd composite) = **39 + 45** (sum of 2 odd composites)

42) **86** = 43 + 43 = 3 + 83 = 7 + 79 = 13 + 73 = 19 + 67 = 43 + 43 = **5 + 81** (sum of 1 prime & 1 odd composite) = **9 + 77** (sum of 2 odd composites)

43) **88** = 5 + 83 = 17 + 71 = 29 + 59 = 41 + 47 = **79 + 9** (sum of 1 prime & 1 odd composite) = **25 + 63** (sum of 2 odd composites)

44) **90** = 7 + 83 = 11 + 79 = 17 + 73 = 19 + 71 = 23 + 67 = 29 + 61 = 31 + 59 = 37 + 53 = 43 + 47 = 1 + 89 = **3 + 87** (sum of 1 prime & 1 odd composite) = **45 + 45** (sum of 2 odd composites)

45) **92** = 3 + 89 = 13 + 79 = 19 + 73 = 31 + 61 = 1 + 91 = **83 + 9** (sum of 1 prime & 1 odd composite) = **15 + 77** (sum of 2 odd composites)

46) **94** = 5 + 89 = 11 + 83 = 23 + 71 = 41 + 53 = 47 + 47 = **7 + 87** (sum of 1 prime & 1 odd composite) = **45 + 49** (sum of 2 odd composites)

47) **96** = 5 + 91 = 7 + 89 = 13 + 83 = 17 + 79 = 23 + 73 = 29 + 67 = 37 + 59 = 43 + 53 = **3 + 93** (sum of 1 prime & 1 odd composite) = **9 + 87** (sum of 2 odd composites)

48) **98** = 7 + 91 = 19 + 79 = 31 + 67 = 37 + 61 = 1 + 97 = **89 + 9** (sum of 1 prime & 1 odd composite) = **49 + 49** (sum of 2 odd composites)

49) **100** = 3 + 97 = 11 + 89 = 17 + 83 = 29 + 71 = 41 + 59 = 47 + 53 = **91 + 9** (sum of 1 prime & 1 odd composite) = **49 + 51** (sum of 2 odd composites)

50) **102** = 5 + 97 = 11 + 91 = 13 + 89 = 19 + 83 = 23 + 79 = 29 + 73 = 31 + 71 = 41 + 61 = 43 + 59 = 1 + 101 = **3 + 99** (sum of 1 prime & 1 odd composite) = **51 + 51** (sum of 2 odd composites)

.
.

(The above is only a partial or incomplete listing of sums of 1 prime & 1 odd composite, and, sums of 2 odd composites, each of which is equal to the sum of 2 primes as well as an even number. For example, in the list of compositions for the even numbers 4 to 102 … above, in Item (48), we could also have other "combinations" such as: **98** = 7 + 91 = 19 + 79 = 31 + 67 = 37 + 61 = 1 + 97 = **25 + 73** (sum of 1 prime & 1 odd composite) = **21 + 77** (sum of 2 odd composites), et al., in Item (49), we could also have other "combinations" such as: **100** = 3 + 97 = 11 + 89 = 17 + 83 = 29 + 71 = 41 + 59 = 47 + 53 = **31 + 69** (sum of 1 prime & 1 odd composite) = **45 + 55** (sum of 2 odd composites), et al., and, in Item (50), we could also have other "combinations" such as: **102** = 5 + 97 = 11 + 91 = 13 + 89 = 19 + 83 = 23 + 79 = 29 + 73 = 31 + 71 = 41 + 61 = 43 + 59 = 1 + 101 = **17 + 85** (sum of 1 prime & 1 odd composite) = **21 + 81** (sum of 2 odd composites), et al. That is, there are more "combinations" than those shown in the above listing.)

In (d) above, in the list of compositions for the 50 consecutive even numbers 4 to 102 …, the even numbers 4, 6, 8 and 10 are only formed through the summing of 2 primes and not at all through the summing of 1 prime and 1 odd composite, or, the summing of 2 odd composites, which are impossibilities here. These sums of 2 primes are present (always present) throughout the whole list of compositions, from 4 right through to 102, while this is not the case for the sums of 1 prime and 1 odd composite, and, the sums of 2 odd composites.

We reason here by the process of elimination, through analyzing the information in (d) above which pertains to the compositions of the 50 consecutive even numbers 4 to 102 … taken from the infinite list of even numbers. We stated at the beginning the following about the even numbers after 2:-

Firstly, every even number after 2 is:
A) The sum of 2 odd numbers.
   (Every odd number is either a prime which is odd or a composite - product of primes which are odd.
   Notably, every prime with the exception of 2 is an odd number.)

Secondly, every even number after 2 is also (the below-mentioned is the logical consequence of (A) above):
1) The sum of 2 primes which are odd.
2) And/or the sum of 1 prime which is odd and 1 odd composite whose prime factors are odd.
3) And/or the sum of 2 odd composites whose prime factors are odd.

Evidently, at least 1 of (1), (2) & (3) above has to be the "atom" or building-block of the even numbers. In (d) above, we observe the following:-
i) All the 50 consecutive even numbers 4 to 102 … in (d) above taken from the infinite list of even

numbers are sums of 2 primes.

ii) It is impossible for each of the even numbers 4, 6, 8 & 10 in (d) above to be the sum of 1 prime which is odd and 1 odd composite whose prime factors are odd.

iii) It is impossible for each of the even numbers 4, 6, 8, 10, 12, 14, 16, 20, 22, 26, 28, 32 & 38 in (d) above to be the sum of 2 odd composites whose prime factors are odd.

It is evident from (i), (ii) & (iii) above that neither (2) nor (3) can be the "atom" or building-block of the even numbers since they are "incomplete". As (1) - the sum of 2 primes which are odd - is "complete", i.e., always present in the 50 consecutive even numbers 4 to 102 ... in (d) above, unlike (2) & (3), it evidently is the "atom" or building-block of the even numbers. That is, every even number after 2 is evidently the sum of 2 primes which are odd. In fact, a distributed computer search completed in 2008 at the University of Aveiro, Portugal, had verified this for all even numbers up to 12×10^{17}, which is not a small list (it is in fact a long, impressive list, obtainable only with the help of modern computer technology). Definitely, due respectively to (ii) & (iii) above, we cannot say that every even number after 2 is the sum of 1 prime which is odd and 1 odd composite whose prime factors are odd, or, every even number after 2 is the sum of 2 odd composites whose prime factors are odd.

By the above lemma and corollary, the infinitudes of the primes, even numbers and odd numbers indeed imply that there are an infinite number of sums of 2 primes which are odd numbers, which are each equal to an even number. As the sums of 2 primes which are odd numbers are evidently the "atoms" or building-blocks of the even numbers, it also implies that they are infinite, since the even numbers are infinite.

Hypothetically, if on the other hand just 1 of the 3 items stated above, primes, even numbers and odd numbers, were finite, the above-said sums of 2 primes which are odd numbers, each of which is equal to an even number, would be finite. The primes, even numbers and odd numbers are evidently intricately linked, with the primes playing the part of building-blocks of both the even and odd numbers through various "combinations" as is described below. However, as the primes, even numbers and odd numbers are intricately linked, the finiteness (or, infinity) of any 1 of them implies the finiteness (or, infinity) of the other 2, and vice versa. These 3 items are evidently "close comrades-in-arm" working together to give special meaning to the integers. As these 3 are all infinite, it indeed implies that there is an infinitude of even numbers which are infinitely the sums of 2 primes that are odd and infinite.

The prime numbers are evidently the atoms or building-blocks of the integers. The integers are either primes (not divisible by other integers except 1) or composites (divisible by other integers, e.g., the prime numbers), and, even (the sums of 2 primes as conjectured by Goldbach) or odd (primes, or, composites whereby they are divisible by prime factors). Therefore, to determine whether the conjecture that every even number (except the number 2) is the sum of 2 primes is true, it would be appropriate to analyze the evident atoms or building-blocks of the even numbers, viz., the prime numbers. For the solution to this conjecture we note that the primes (vide Euclid's proof) and the even numbers are infinite, which implies that this conjecture should be true.

We here analyze the "behavior" of the first 2,400 consecutive prime numbers (divided into 12 batches of consecutive primes, each subsequent batch with an increment of 200 primes), leaving out 2 (because it is an even prime) and commencing with 3, which is the $2^{nd.}$ consecutive prime, the latter to be the first prime in our list of 2,400 consecutive primes (3 to 21,391), as follows:-

(1) <u>200 Consecutive Primes From 3 To 1,229</u>
   (a) Even numbers (obtained by summing of 2 primes) = 6 to 2,458
   (b) No. of even numbers = 1,227
   (c) No. of primes = 200
   (d) Average no. of even numbers "generated" by each of these 200 consecutive primes = 1,227 ÷ 200 = **6.14**
   (e) No. of summings of 2 primes/permutations (3 + 3, 3 + 5, 3 + 7, 3 + 11, ...et al.) for these 200 primes = 200 x 200 = 40,000
   (f) Average no. of summings of 2 primes/permutations for each of the 1,227 even numbers = 40,000 ÷ 1,227 = **32.60**

(2) <u>400 Consecutive Primes From 3 To 2,749</u>
   (a) Even numbers (obtained by summing of 2 primes) = 6 to 5,498
   (b) No. of even numbers = 2,747
   (c) No. of primes = 400
   (d) Average no. of even numbers "generated" by each of these 400 consecutive primes = 2,747 ÷ 400 = **6.87**
   (e) No. of summings of 2 primes/permutations (3 + 3, 3 + 5, 3 + 7, 3 + 11, ...et al.) for these 400 primes = 400 x 400 = 160,000
   (f) Average no. of summings of 2 primes/permutations for each of the 2,747 even numbers = 160,000 ÷ 2,747 = **58.25**

(3) <u>600 Consecutive Primes From 3 To 4,421</u>
   (a) Even numbers (obtained by summing of 2 primes) = 6 to 8,842
   (b) No. of even numbers = 4,419
   (c) No. of primes = 600
   (d) Average no. of even numbers "generated" by each of these 600 consecutive primes = 4,419 ÷ 600

= **7.37**

(e) No. of summings of 2 primes/permutations (3 + 3, 3 + 5, 3 + 7, 3 + 11, ...et al.) for these 600 primes = 600 x 600 = 360,000

(f) Average no. of summings of 2 primes/permutations for each of the 4,419 even numbers = 360,000 ÷ 4,419 = **81.47**

(4) 800 Consecutive Primes From 3 To 6,143
  (a) Even numbers (obtained by summing of 2 primes) = 6 to 12,286
  (b) No. of even numbers = 6,141
  (c) No. of primes = 800
  (d) Average no. of even numbers "generated" by each of these 800 consecutive primes = 6,141 ÷ 800 = **7.68**
  (e) No. of summings of 2 primes/permutations (3 + 3, 3 + 5, 3 + 7, 3 + 11, ...et al.) for these 800 primes = 800 x 800 = 640,000
  (f) Average no. of summings of 2 primes/permutations for each of the 6,141 even numbers = 640,000 ÷ 6,141 = **104.22**

(5) 1,000 Consecutive Primes From 3 To 7,927
  (a) Even numbers (obtained by summing of 2 primes) = 6 to 15,854
  (b) No. of even numbers = 7,925
  (c) No. of primes = 1,000
  (d) Average no. of even numbers "generated" by each of these 1,000 consecutive primes = 7,925 ÷ 1,000 = **7.93**
  (e) No. of summings of 2 primes/permutations (3 + 3, 3 + 5, 3 + 7, 3 + 11, ...et al.) for these 1,000 primes = 1,000 x 1,000 = 1,000,000
  (f) Average no. of summings of 2 primes/permutations for each of the 7,925 even numbers = 1,000,000 ÷ 7,925 = **126.18**

(6) 1,200 Consecutive Primes From 3 To 9,739
  (a) Even numbers (obtained by summing of 2 primes) = 6 to 19,478
  (b) No. of even numbers = 9,737
  (c) No. of primes = 1,200
  (d) Average no. of even numbers "generated" by each of these 1,200 consecutive primes = 9,737 ÷ 1,200 = **8.11**
  (e) No. of summings of 2 primes/permutations (3 + 3, 3 + 5, 3 + 7, 3 + 11, ...et al.) for these 1,200 primes = 1,200 x 1,200 = 1,440,000
  (f) Average no. of summings of 2 primes/permutations for each of the 9,737 even numbers =

$1,440,000 \div 9,737 = \mathbf{147.89}$

(7) <u>1,400 Consecutive Primes From 3 To 11,677</u>
    (a) Even numbers (obtained by summing of 2 primes) = 6 to 23,354
    (b) No. of even numbers = 11,675
    (c) No. of primes = 1,400
    (d) Average no. of even numbers "generated" by each of these 1,400 consecutive primes = 11,675 ÷ 1,400 = **8.34**
    (e) No. of summings of 2 primes/permutations (3 + 3, 3 + 5, 3 + 7, 3 + 11, ……et al.) for these 1,400 primes = 1,400 x 1,400 = 1,960,000
    (f) Average no. of summings of 2 primes/permutations for each of the 11,675 even numbers = 1,960,000 ÷ 11,675 = **167.88**

(8) <u>1,600 Consecutive Primes From 3 To 13,513</u>
    (a) Even numbers (obtained by summing of 2 primes) = 6 to 27,026
    (b) No. of even numbers = 13,511
    (c) No. of primes = 1,600
    (d) Average no. of even numbers "generated" by each of these 1,600 consecutive primes = 13,511 ÷ 1,600 = **8.44**
    (e) No. of summings of 2 primes/permutations (3 + 3, 3 + 5, 3 + 7, 3 + 11, ……et al.) for these 1,600 primes = 1,600 x 1,600 = 2,560,000
    (f) Average no. of summings of 2 primes/permutations for each of the 13,511 even numbers = 2,560,000 ÷ 13,511 = **189.48**

(9) <u>1,800 Consecutive Primes From 3 To 15,413</u>
    (a) Even numbers (obtained by summing of 2 primes) = 6 to 30,826
    (b) No. of even numbers = 15,411
    (c) No. of primes = 1,800
    (d) Average no. of even numbers "generated" by each of these 1,800 consecutive primes = 15,411 ÷ 1,800 = **8.56**
    (e) No. of summings of 2 primes/permutations (3 + 3, 3 + 5, 3 + 7, 3 + 11, ……et al.) for these 1,800 primes = 1,800 x 1,800 = 3,240,000
    (f) Average no. of summings of 2 primes/permutations for each of the 15,411 even numbers = 3,240,000 ÷ 15,411 = **210.24**

(10) <u>2,000 Consecutive Primes From 3 To 17,393</u>
    (a) Even numbers (obtained by summing of 2 primes) = 6 to 34,786

(b) No. of even numbers = 17,391

(c) No. of primes = 2,000

(d) Average no. of even numbers "generated" by each of these 2,000 consecutive primes = 17,391 ÷ 2,000 = **8.70**

(e) No. of summings of 2 primes/permutations (3 + 3, 3 + 5, 3 + 7, 3 + 11, ……et al.) for these 2,000 primes = 2,000 x 2,000 = 4,000,000

(f) Average no. of summings of 2 primes/permutations for each of the 17,391 even numbers = 4,000,000 ÷ 17,391 = **230.00**

(11) 2,200 Consecutive Primes From 3 To 19,427

(a) Even numbers (obtained by summing of 2 primes) = 6 to 38,854

(b) No. of even numbers = 19,425

(c) No. of primes = 2,200

(d) Average no. of even numbers "generated" by each of these 2,200 consecutive primes = 19,425 ÷ 2,200 = **8.83**

(e) No. of summings of 2 primes/permutations (3 + 3, 3 + 5, 3 + 7, 3 + 11, ……et al.) for these 2,200 primes = 2,200 x 2,200 = 4,840,000

(f) Average no. of summings of 2 primes/permutations for each of the 19,425 even numbers = 4,840,000 ÷ 19,425 = **249.16**

(12) 2,400 Consecutive Primes From 3 To 21,391

(a) Even numbers (obtained by summing of 2 primes) = 6 to 42,782

(b) No. of even numbers = 21,389

(c) No. of primes = 2,400

(d) Average no. of even numbers "generated" by each of these 2,400 consecutive primes = 21,389 ÷ 2,400 = **8.91**

(e) No. of summings of 2 primes/permutations (3 + 3, 3 + 5, 3 + 7, 3 + 11, ……et al.) for these 2,400 primes = 2,400 x 2,400 = 5,760,000

(f) Average no. of summings of 2 primes/permutations for each of the 21,389 even numbers = 5,760,000 ÷ 21,389 = **269.30**

There would evidently be more and more profuse repetitions and overlaps of the even numbers "generated" by the primes the higher up the infinite list of prime numbers we go, which is significant. (For a better insight of this, refer to Appendix 1 and Appendix 2.)

We compare all the (d)s and (f)s in (1) to (12) above, which is as follows:-

(d) Average no. of even numbers "generated" by each of the consecutive primes in (1) to (12) above, as follows according to the listings (1) to (12):

(1) **6.14**
(2) **6.87**
(3) **7.37**
(4) **7.68**
(5) **7.93**
(6) **8.11**
(7) **8.34**
(8) **8.44**
(9) **8.56**
(10) **8.70**
(11) **8.83**
(12) **8.91**

(f) Average no. of summings of 2 primes/permutations for each of the even numbers in (1) to (12) above, as follows according to the listings (1) to (12):

(1) **32.60**
(2) **58.25**
(3) **81.47**
(4) **104.22**
(5) **126.18**
(6) **147.89**
(7) **167.88**
(8) **189.48**
(9) **210.24**
(10) **230.00**
(11) **249.16**
(12) **269.30**

The following is evident from the above information:-

(A): (d) Average no. of even numbers "generated" by each of the consecutive primes in the above 12 listings increases continually all the way from the list: (1) 200 Consecutive Primes From 3 To 1,229 to the list: (12) 2,400 Consecutive Primes From 3 To 21,391, from **6.14** even numbers per prime number in List (1) to **8.91** even numbers per prime number in List (12).

(B): (f) Average no. of summings of 2 primes/permutations for each of the even numbers in the above 12 listings increases continually all the way from the list: (1) 200 Consecutive Primes From 3 To 1,229 to the list: (12) 2,400 Consecutive Primes From 3 To 21,391, from **32.60** number of summings of 2 primes/permutations per even number in List (1) to **269.30** number of summings of 2 primes/permutations per even number in List (12).

Lemma:
According to the principle of complete induction in set theory, if a set of natural numbers contains 1 and, for each n, it contains $n + 1$ whenever it contains all numbers less than $n + 1$, then it must contain every natural number, e.g., complete induction proves that every natural number is a product of primes.

By induction, we now deduce the following:

The larger the list of consecutive primes becomes, the greater would be the average number of even numbers "generated" by each of the primes in the list of consecutive primes (inferred from (A) above).

The larger the list of consecutive primes becomes, the greater would be the average number of summings of 2 primes/permutations for each of the even numbers in the infinite list of even numbers (inferred from (B) above).

Furthermore, the Goldbach conjecture had been tested and found to be correct for every even number up to 12×10^{17}, which is not a small list, by a distributed computer search carried out at the University of Aveiro, Portugal, in 2008.

As the primes and the even numbers are infinite, by the above lemma and all the above deductions and information, it could be inferred that the increases stated in (A) and (B) above, with the even numbers each being the sum of 2 primes, continue to infinity, *i.e., the Goldbach conjecture becomes stronger and stronger the higher up the infinite list of prime numbers/even numbers we go - all the way to infinity.*

Next, we resort to the proof by contradiction. The above deduction would be reversed if, e.g., the following takes place (which is the reversal of the above-mentioned information):

(A): (d) Average no. of even numbers "generated" by each of the consecutive primes in the above 12 listings decreases continually all the way from the list: (1) 200 Consecutive Primes From 3 To 1,229 to the list: (12) 2,400 Consecutive Primes From 3 To 21,391, from **8.91** even numbers per prime number in List (1) to **6.14** even numbers per prime number in List (12).

(B): (f) Average no. of summings of 2 primes/permutations for each of the even numbers in the above 12 listings decreases continually all the way from the list: (1) 200 Consecutive Primes From 3 To 1,229 to the list: (12) 2,400 Consecutive Primes From 3 To 21,391, from **269.30** number of summings of 2 primes/permutations per even number in List (1) to **32.60** number of summings of 2 primes/permutations per even number in List (12).

If this reversed state happens, the implication is that there would reach a point when there are no more batches of 2 prime numbers summing together to form even numbers, in which case the Goldbach conjecture would be false. Evidently this would happen when the prime numbers are finite. As the prime numbers are infinite (as Euclid had proved long ago) this would never happen.

Since the above information indicate otherwise, and, the prime numbers are infinite, we accept the above induction/deduction and infer that the Goldbach conjecture could not be false, i.e., the Goldbach conjecture is true, and, every even number (except 2) is indeed the sum of 2 prime numbers. This concludes the proof by contradiction.

We take a look at the following example to see how effectively the primes "generate" new even numbers in accordance with the Goldbach conjecture:-

Density Of New Even Numbers "Generated" (*See Appendix 1 For Example Of Computation Method*)

(a) Set Of Integers, 51 To 100, With 10 Primes Within It = 5 New Even Nos. Per Prime No.
  (No. Of New Even Nos. "Generated" = 50. No. Of Primes = 10.)

(b) Set Of Integers, 101 To 150, With 10 Primes Within It = 5.2 New Even Nos. Per Prime No.
  (No. Of New Even Nos. "Generated" = 52. No. Of Primes = 10.)

(c) Set Of Integers, 151 To 200, With 11 Primes Within It = 4.55 New Even Nos. Per Prime No.
  (No. Of New Even Nos. "Generated" = 50. No. Of Primes = 11.)

(d) Set Of Integers, 201 To 250, With 7 Primes Within It = 6 New Even Nos. Per Prime No.
  (No. Of New Even Nos. "Generated" = 42. No. Of Primes = 7.)

(e) Set Of Integers, 251 To 300, With 9 Primes Within It = 5.78 New Even Nos. Per Prime No.
  (No. Of New Even Nos. "Generated" = 52. No. Of Primes = 9.)

(f) Set Of Integers, 301 To 350, With 8 Primes Within It = 7 New Even Nos. Per Prime No. (No. Of New Even Nos. "Generated" = 56. No. Of Primes = 8.)

(g) Set Of Integers, 351 To 400, With 8 Primes Within It = 6 New Even Nos. Per Prime No. (No. Of New Even Nos. "Generated" = 48. No. Of Primes = 8.)

(h) Set Of Integers, 401 To 450, With 9 Primes Within It = 5.78 New Even Nos. Per Prime No. (No. Of New Even Nos. "Generated" = 52. No. Of Primes = 9.)

(i) Set Of Integers, 451 To 500, With 8 Primes Within It = 6.25 New Even Nos. Per Prime No. (No. Of New Even Nos. "Generated" = 50. No. Of Primes = 8.)

(j) Set Of Integers, 501 To 550, With 6 Primes Within It = 8 New Even Nos. Per Prime No. (No. Of New Even Nos. "Generated" = 48. No. Of Primes = 6.)

(k) Set Of Integers, 551 To 600, With 8 Primes Within It = 6.5 New Even Nos. Per Prime No. (No. Of New Even Nos. "Generated" = 52. No. Of Primes = 8.)

(l) Set Of Integers, 601 To 650, With 9 Primes Within It = 5.33 New Even Nos. Per Prime No. (No. Of New Even Nos. "Generated" = 48. No. Of Primes = 9.)

(m) Set Of Integers, 651 To 700, With 7 Primes Within It = 6.29 New Even Nos. Per Prime No. (No. Of New Even Nos. "Generated" = 44. No. Of Primes = 7.)

(n) Set Of Integers, 701 To 750, With 7 Primes Within It = 7.43 New Even Nos. Per Prime No. (No. Of New Even Nos. "Generated" = 52. No. Of Primes = 7.)

(o) Set Of Integers, 751 To 800, With 7 Primes Within It = 7.71 New Even Nos. Per Prime No. (No. Of New Even Nos. "Generated" = 54. No. Of Primes = 7.)

(p) Set Of Integers, 801 To 850, With 7 Primes Within It = 6 New Even Nos. Per Prime No. (No. Of New Even Nos. "Generated" = 42. No. Of Primes = 7.)

(q) Set Of Integers, 851 To 900, With 8 Primes Within It = 6 New Even Nos. Per Prime No. (No. Of New Even Nos. "Generated" = 48. No. Of Primes = 8.)

(r) Set Of Integers, 901 To 950, With 7 Primes Within It = 8.57 New Even Nos. Per Prime No. (No. Of New Even Nos. "Generated" = 60. No. Of Primes = 7.)

(s) Set Of Integers, 951 To 1,000, With 7 Primes Within It = 7.14 New Even Nos. Per Prime No. (No. Of New Even Nos. "Generated" = 50. No. Of Primes = 7.)

(t) Set Of Integers, 1,001 To 1,050, With 8 Primes Within It = 6.5 New Even Nos. Prime No.
(No. Of New Even Nos. "Generated" = 52. No. Of Primes = 8.)

(u) Set Of Integers, 1,051 To 1,100, With 8 Primes Within It = 6 New Even Nos. Per Prime No.
(No. Of New Even Nos. "Generated" = 48. No. Of Primes = 8.)

(v) Set Of Integers, 1,101 To 1,150, With 5 Primes Within It = 6.4 New Even Nos. Per Prime No.
(No. Of New Even Nos. "Generated" = 32. No. Of Primes = 5.)

(w) Set Of Integers, 1,151 To 1,200, With 7 Primes Within It = **9.14 New Even Nos. Per Prime No.**
(No. Of New Even Nos. "Generated" = 64. No. Of Primes = 7.)

(x) Set Of Integers, 1,201 To 1,250, With 8 Primes Within It = 7 New Even Nos. Per Prime No.
(No. Of New Even Nos. "Generated" = 56. No. Of Primes = 8.)

Average Density For The Above 24 Items ((a) To (x)) = 155.54 ÷ 24 = 6.48 New Even Nos. Per Prime No.

Maximum Density = 9.14 New Even Nos. Per Prime No. (No. Of New Even Nos. "Generated" = 64. No. Of Primes = 7.)

Minimum Density = 4.55 New Even Nos. Per Prime No. (No. Of New Even Nos. "Generated" = 50. No. Of Primes = 11.)

Such a "profuse generation" of "regular batches" of even numbers by the prime numbers is significant and lends further support to the validity of the Goldbach conjecture.

There is further proof which is obtainable by analyzing a number of even numbers, e.g., we could split a group of 240 even consecutive numbers, from 4 to 482, into 8 equal batches (30 even numbers per batch) and analyze the batches; this would corroborate the fact that the infinite quantity of primes would "generate" a regular, continuous (*without breaks or gaps*) and infinite list of even numbers. The density of distribution or prime additions/combinations per even number evidently become greater and greater the higher up the infinite list of the even numbers we go, i.e., *the Goldbach conjecture evidently becomes stronger and stronger the higher up the infinite list of the even numbers we go*. This pattern is significant and is discernable in the following example:-

(1) 1st. Batch Of 30 Even Numbers (4 To 62)   (*See Appendix 2 For Example Of Computation Method*)
a) Maximum No. Of Prime Additions/Combinations Per Even Number = **5**
b) Minimum No. Of Prime Additions/Combinations Per Even Number = **1**

c) Density Of Distribution = Average Prime Additions/Combinations Per Even Number = **2.77** Prime Additions/Combinations Per Even Number

(2) 2 nd. Batch Of 30 Even Numbers (64 To 122)

    a) Maximum No. Of Prime Additions/Combinations Per Even Number = 14
    b) Minimum No. Of Prime Additions/Combinations Per Even Number = 2
    c) Density Of Distribution = Average Prime Additions/Combinations Per Even Number = **6.1** Prime Additions/Combinations Per Even Number
    d) Percentage Increase In Density Of Distribution = (6.1 - 2.77) ÷ 2.77 x 100% = 120.22%

(3) 3 rd. Batch Of 30 Even Numbers (124 To 182)

    a) Maximum No. Of Prime Additions/Combinations Per Even Number = 16
    b) Minimum No. Of Prime Additions/Combinations Per Even Number = 4
    c) Density Of Distribution = Average Prime Additions/Combinations Per Even Number = **9.07** Prime Additions/Combinations Per Even Number
    d) Percentage Increase In Density Of Distribution = (9.07 - 6.1) ÷ 6.1 x 100% = 48.69%

(4) 4 th. Batch Of 30 Even Numbers (184 To 242)

    a) Maximum No. Of Prime Additions/Combinations Per Even Number = 22
    b) Minimum No. Of Prime Additions/Combinations Per Even Number = 5
    c) Density Of Distribution = Average Prime Additions/Combinations Per Even Number = **10.53** Prime Additions/Combinations Per Even Number
    d) Percentage Increase In Density Of Distribution = (10.53 - 9.07) ÷ 9.07 x 100% = 16.1%

(5) 5 th. Batch Of 30 Even Numbers (244 To 302)

    a) Maximum No. Of Prime Additions/Combinations Per Even Number = 21
    b) Minimum No. Of Prime Additions/Combinations Per Even Number = 7
    c) Density Of Distribution = Average Prime Additions/Combinations Per Even Number = **12.37** Prime Additions/Combinations Per Even Number
    d) Percentage Increase In Density Of Distribution = (12.37 - 10.53) ÷ 10.53 x 100% = 17.47%

(6) 6 th. Batch Of 30 Even Numbers (304 To 362)

    a) Maximum No. Of Prime Additions/Combinations Per Even Number = 27
    b) Minimum No. Of Prime Additions/Combinations Per Even Number = 7
    c) Density Of Distribution = Average Prime Additions/Combinations Per Even Number = **13.77** Prime Additions/Combinations Per Even Number
    d) Percentage Increase In Density Of Distribution = (13.77 - 12.37) ÷ 12.37 x 100% = 11.32%

(7) 7 th. Batch Of 30 Even Numbers (364 To 422)

    a) Maximum No. Of Prime Additions/Combinations Per Even Number = 30
    b) Minimum No. Of Prime Additions/Combinations Per Even Number = 7
    c) Density Of Distribution = Average Prime Additions/Combinations Per Even Number = **15.23** Prime Additions/Combinations Per Even Number
    d) Percentage Increase In Density Of Distribution = (15.23 - 13.77) ÷ 13.77 x 100% = 10.6%

(8) 8 th. Batch Of 30 Even Numbers (424 To 482)

    a) Maximum No. Of Prime Additions/Combinations Per Even Number = 30
    b) Minimum No. Of Prime Additions/Combinations Per Even Number = 9
    c) Density Of Distribution = Average Prime Additions/Combinations Per Even Number = **16.93** Prime Additions/Combinations Per Even Number
    d) Percentage Increase In Density Of Distribution = (16.93 - 15.23) ÷ 15.23 x 100% = 11.16%

The Density Of Distribution is expected to increase to infinity, though the Percentage Increase In Density Of Distribution is expected to thin out towards infinity - it could be seen above to increase from 2.77 prime additions/combinations per even number for batch of even numbers, 4 to 62, all the way up to 16.93 prime additions/combinations per even number for batch of even numbers, 424 to 482. This is nevertheless significant evidence that lends support to the validity of the Goldbach conjecture. Also, the Maximum No. Of Prime Additions/Combinations Per Even Number and the Minimum No. Of Prime Additions/Combinations Per Even Number could be seen to range from 5 and 1 respectively for batch of even numbers, 4 to 62, to 30 and 9 respectively for batch of even numbers, 424 to 482. This trend of "upward increase" of the (maximum and minimum) numbers of prime additions/combinations for each even number implies that at some points toward infinity the numbers of prime additions/combinations for each even number could be thousands, millions, billions, trillions, and more, if only we have the computing power to compute/check such prime additions/combinations (this again indicates that *the Goldbach conjecture becomes evidently stronger and stronger the higher up the infinite list of the even numbers we go*). This is significant too and is also evidence that lends support to the validity of the Goldbach conjecture. By the infinitude of the primes (vide Euclid's proof) and even numbers, these "patterns", as described here, would be there all the way to infinity, which would be in accordance with the Goldbach conjecture.

The following evidence would further affirm the validity of the Goldbach conjecture:-

1) 10 consecutive primes, commencing from the odd prime 3, would give rise to 10 x 10, or, 100 sums of 2 primes/partitions/permutations, but less than 100 different even numbers, with many repetitions/overlaps (e.g., for these first 10 consecutive primes 3, 5, 7, 11, 13, 17, 19, 23, 29 & 31, 10 = 3 + 7 = 5 + 5 (2 partitions/permutations), 22 = 3 + 19 = 5 + 17 = 11 + 11 (3 partitions/permutations), &, 34 = 3 + 31 = 5 + 29 = 11 + 23 = 17 + 17 (4 partitions/permutations)).

2) 20 consecutive primes, commencing from the odd prime 3, (increase of **100%** in no. of consecutive primes compared to (1) above) would give rise to 20 x 20, or, 400 sums of 2 primes/partitions/permutations (increase of **300%** in no. of sums of 2 primes/partitions/permutations compared to (1) above), but less than 400 different even numbers, with many repetitions/overlaps.

3) 30 consecutive primes, commencing from the odd prime 3, (increase of **200%** in no. of consecutive primes compared to (1) above) would give rise to 30 x 30, or, 900 sums of 2 primes/partitions/permutations (increase of **800%** in no. of sums of 2 primes/partitions/permutations compared to (1) above), but less than 900 different even numbers, with many repetitions/overlaps.

4) 40 consecutive primes, commencing from the odd prime 3, (increase of **300%** in no. of consecutive primes compared to (1) above) would give rise to 40 x 40, or, 1,600 sums of 2 primes/partitions/permutations (increase of **1,500%** in no. of sums of 2 primes/partitions/permutations compared to (1) above), but less than 1,600 different even numbers, with many repetitions/overlaps.

5) 50 consecutive primes, commencing from the odd prime 3, (increase of **400%** in no. of consecutive primes compared to (1) above) would give rise to 50 x 50, or, 2,500 sums of 2 primes/partitions/permutations (increase of **2,400%** in no. of sums of 2 primes/partitions/permutations compared to (1) above), but less than 2,500 different even numbers, with many repetitions/overlaps.

6) 60 consecutive primes, commencing from the odd prime 3, (increase of **500%** in no. of

consecutive primes compared to (1) above) would give rise to 60 x 60, or, 3,600 sums of 2 primes/partitions/permutations (increase of **3,500%** in no. of sums of 2 primes/partitions/permutations compared to (1) above), but less than 3,600 different even numbers, with many repetitions/overlaps.

7) 70 consecutive primes, commencing from the odd prime 3, (increase of **600%** in no. of consecutive primes compared to (1) above) would give rise to 70 x 70, or, 4,900 sums of 2 primes/partitions/permutations (increase of **4,800%** in no. of sums of 2 primes/partitions/permutations compared to (1) above), but less than 4,900 different even numbers, with many repetitions/overlaps.

8) 80 consecutive primes, commencing from the odd prime 3, (increase of **700%** in no. of consecutive primes compared to (1) above) would give rise to 80 x 80, or, 6,400 sums of 2 primes/partitions/permutations (increase of **6,300%** in no. of sums of 2 primes/partitions/permutations compared to (1) above), but less than 6,400 different even numbers, with many repetitions/overlaps.

9) 90 consecutive primes, commencing from the odd prime 3, (increase of **800%** in no. of consecutive primes compared to (1) above) would give rise to 90 x 90, or, 8,100 sums of 2 primes/partitions/permutations (increase of **8,000%** in no. of sums of 2 primes/partitions/permutations compared to (1) above), but less than 8,100 different even numbers, with many repetitions/overlaps.

10) 100 consecutive primes, commencing from the odd prime 3, (increase of **900%** in no. of consecutive primes compared to (1) above) would give rise to 100 x 100, or, 10,000 sums of 2 primes/partitions/permutations (increase of **9,900%** in no. of sums of 2 primes/partitions/permutations compared to (1) above), but less than 10,000 different even numbers, with many repetitions/overlaps.

.

.

.

.

The following is evident from the above:-

1) The 1st. marginal increase of **100%** in no. of consecutive primes (increase of 200% - increase of 100%) results in marginal increase of **500%** in no. of sums of 2 primes/partitions/permutations (increase of 800% - increase of 300%).

2) The 2nd. marginal increase of **100%** in no. of consecutive primes (increase of 300% - increase of 200%) results in marginal increase of **700%** in no. of sums of 2 primes/partitions/permutations (increase of 1,500% - increase of 800%).

3) The 3rd. marginal increase of **100%** in no. of consecutive primes (increase of 400% - increase of 300%) results in marginal increase of **900%** in no. of sums of 2 primes/partitions/permutations (increase of 2,400% - increase of 1,500%).

4) The 4th. marginal increase of **100%** in no. of consecutive primes (increase of 500% - increase of 400%) results in marginal increase of **1,100%** in no. of sums of 2 primes/partitions/permutations (increase of 3,500% - increase of 2,400%).

5) The 5th. marginal increase of **100%** in no. of consecutive primes (increase of 600% - increase of 500%) results in marginal increase of **1,300%** in no. of sums of 2 primes/partitions/permutations (increase of 4,800% - increase of 3,500%).

6) The 6th. marginal increase of **100%** in no. of consecutive primes (increase of 700% - increase of 600%) results in marginal increase of **1,500%** in no. of sums of 2 primes/partitions/permutations (increase of 6,300% - increase of 4,800%).

7) The 7th. marginal increase of **100%** in no. of consecutive primes (increase of 800% - increase of 700%) results in marginal increase of **1,700%** in no. of sums of 2 primes/partitions/permutations (increase of 8,000% - increase of 6,300%).

8) The 8th. marginal increase of **100%** in no. of consecutive primes (increase of 900% - increase of 800%) results in marginal increase of **1,900%** in no. of sums of 2 primes/partitions/permutations (increase of 9,900% - increase of 8,000%).

.
.
.
.

(1) to (8) above show that while the marginal increase in no. of consecutive primes remains constant at 100% from (1) to (8), the marginal increase in no. of sums of 2 primes/partitions/permutations goes up progressively from 500% in (1) to 1,900% in (8). It is evident here that the higher up the infinite list of primes we go, the more "overwhelming" or dense the (one-to-one) combinations of primes (i.e., sums of 2 primes, in the formation of even numbers) would become, the number of permutations of the combinations of primes tending towards infinity (with the infinity of the prime numbers). *In other words, the Goldbach conjecture becomes stronger and stronger the higher up the infinite list of prime numbers/even numbers we go.* The infinitude of the prime numbers (vide Euclid's proof) and even numbers would hence imply the validity of the Goldbach conjecture.

The prime number theorem, which had been proven, states that the limit of the quotient of the 2 functions $\pi(n)$ and $n/\log n$ as n approaches infinity is 1, which is expressed by the formula:

$$\lim_{n \to \infty} \pi(n)/(n/\log n) = 1$$

where $\pi(n)$ is approximately equal to $(n/\log n)$

The function $\pi(n)$ represents the number of primes less than or equal to the number n. This function measures the distribution of the prime numbers. With it, we compute the ratio $n/\pi(n)$ which says what fraction of the numbers up to a given point are primes. (It is actually the reciprocal of this fraction.) The following is the result of a computation:-

n	*π(n)*	*n/π(n)*
10	4 (a)	2.5
100	25 (b)	4.0
1,000	168 (c)	6.0
10,000	1,229 (d)	8.1
100,000	9,592 (e)	10.4
1,000,000	78,498 (f)	12.7
10,000,000	664,579 (g)	15.0
100,000,000	5,761,455 (h)	17.4
1,000,000,000	50,847,534 (i)	19.7
10,000,000,000	455,052,512 (j)	22.0

It is noticeable that as one moves from 1 power of 10 to the next, the ratio $n/\pi(n)$ increases by about 2.3, e.g., 22.0 - 19.7 = 2.3. As $\log_e 10 = 2.30258$ ..., we may thus regard $\pi(n)$ as approximately equal to $n/\log n$.

We have the following partitions with the primes described in the "$\pi(n)$" column above:-

1) With (a) above, we have the following "prime + prime = even number" combinations:

a) prime a + prime a: 4 x 4 "prime + prime" combinations
b) prime a + prime b: 4 x 25 "prime + prime" combinations
c) prime a + prime c: 4 x 168 "prime + prime" combinations
d) prime a + prime d: 4 x 1,229 "prime + prime" combinations
e) prime a + prime e: 4 x 9,592 "prime + prime" combinations

f) prime a + prime f: 4 x 78,498 "prime + prime" combinations
g) prime a + prime g: 4 x 664,579 "prime + prime" combinations
h) prime a + prime h: 4 x 5,761,455 "prime + prime" combinations
i) prime a + prime i: 4 x 50,847,534 "prime + prime" combinations
j) prime a + prime j: 4 x 455,052,512 "prime + prime" combinations

For example, for (j) above, a prime described in (a) in the "$\pi(n)$" column above plus a prime described in (j) in the "$\pi(n)$" column above give an even number, and there are 4 x 455,052,512 such "prime + prime = even number" combinations.

2) With (b) above, we have the following "prime + prime = even number" combinations:

a) prime b + prime a: 25 x 4 "prime + prime" combinations
b) prime b + prime b: 25 x 25 "prime + prime" combinations
c) prime b + prime c: 25 x 168 "prime + prime" combinations
d) prime b + prime d: 25 x 1,229 "prime + prime" combinations
e) prime b + prime e: 25 x 9,592 "prime + prime" combinations
f) prime b + prime f: 25 x 78,498 "prime + prime" combinations
g) prime b + prime g: 25 x 664,579 "prime + prime" combinations
h) prime b + prime h: 25 x 5,761,455 "prime + prime" combinations
i) prime b + prime i: 25 x 50,847,534 "prime + prime" combinations
j) prime b + prime j: 25 x 455,052,512 "prime + prime" combinations

3) With (c) above, we have the following "prime + prime = even number" combinations:

a) prime c + prime a: 168 x 4 "prime + prime" combinations
b) prime c + prime b: 168 x 25 "prime + prime" combinations
c) prime c + prime c: 168 x 168 "prime + prime" combinations
d) prime c + prime d: 168 x 1,229 "prime + prime" combinations
e) prime c + prime e: 168 x 9,592 "prime + prime" combinations
f) prime c + prime f: 168 x 78,498 "prime + prime" combinations
g) prime c + prime g: 168 x 664,579 "prime + prime" combinations
h) prime c + prime h: 168 x 5,761,455 "prime + prime" combinations
i) prime c + prime i: 168 x 50,847,534 "prime + prime" combinations
j) prime c + prime j: 168 x 455,052,512 "prime + prime" combinations

4) With (d) above, we have the following "prime + prime = even number" combinations:

a) prime d + prime a: 1,229 x 4 "prime + prime" combinations
b) prime d + prime b: 1,229 x 25 "prime + prime" combinations
c) prime d + prime c: 1,229 x 168 "prime + prime" combinations
d) prime d + prime d: 1,229 x 1,229 "prime + prime" combinations
e) prime d + prime e: 1,229 x 9,592 "prime + prime" combinations
f) prime d + prime f: 1,229 x 78,498 "prime + prime" combinations
g) prime d + prime g: 1,229 x 664,579 "prime + prime" combinations
h) prime d + prime h: 1,229 x 5,761,455 "prime + prime" combinations
i) prime d + prime i: 1,229 x 50,847,534 "prime + prime" combinations
j) prime d + prime j: 1,229 x 455,052,512 "prime + prime" combinations

5) With (e) above, we have the following "prime + prime = even number" combinations:

a) prime e + prime a: 9,592 x 4 "prime + prime" combinations
b) prime e + prime b: 9,592 x 25 "prime + prime" combinations
c) prime e + prime c: 9,592 x 168 "prime + prime" combinations
d) prime e + prime d: 9,592 x 1,229 "prime + prime" combinations
e) prime e + prime e: 9,592 x 9,592 "prime + prime" combinations
f) prime e + prime f: 9,592 x 78,498 "prime + prime" combinations
g) prime e + prime g: 9,592 x 664,579 "prime + prime" combinations
h) prime e + prime h: 9,592 x 5,761,455 "prime + prime" combinations
i) prime e + prime i: 9,592 x 50,847,534 "prime + prime" combinations
j) prime e + prime j: 9,592 x 455,052,512 "prime + prime" combinations

6) With (f) above, we have the following "prime + prime = even number" combinations:

a) prime f + prime a: 78,498 x 4 "prime + prime" combinations
b) prime f + prime b: 78,498 x 25 "prime + prime" combinations
c) prime f + prime c: 78,498 x 168 "prime + prime" combinations
d) prime f + prime d: 78,498 x 1,229 "prime + prime" combinations
e) prime f + prime e: 78,498 x 9,592 "prime + prime" combinations
f) prime f + prime f: 78,498 x 78,498 "prime + prime" combinations

g) prime f + prime g: 78,498 x 664,579 "prime + prime" combinations
h) prime f + prime h: 78,498 x 5,761,455 "prime + prime" combinations
i) prime f + prime i: 78,498 x 50,847,534 "prime + prime" combinations
j) prime f + prime j: 78,498 x 455,052,512 "prime + prime" combinations

7) With (g) above, we have the following "prime + prime = even number" combinations:

a) prime g + prime a: 664,579 x 4 "prime + prime" combinations
b) prime g + prime b: 664,579 x 25 "prime + prime" combinations
c) prime g + prime c: 664,579 x 168 "prime + prime" combinations
d) prime g + prime d: 664,579 x 1,229 "prime + prime" combinations
e) prime g + prime e: 664,579 x 9,592 "prime + prime" combinations
f) prime g + prime f: 664,579 x 78,498 "prime + prime" combinations
g) prime g + prime g: 664,579 x 664,579 "prime + prime" combinations
h) prime g + prime h: 664,579 x 5,761,455 "prime + prime" combinations
i) prime g + prime i: 664,579 x 50,847,534 "prime + prime" combinations
j) prime g + prime j: 664,579 x 455,052,512 "prime + prime" combinations

8) With (h) above, we have the following "prime + prime = even number" combinations:

a) prime h + prime a: 5,761,455 x 4 "prime + prime" combinations
b) prime h + prime b: 5,761,455 x 25 "prime + prime" combinations
c) prime h + prime c: 5,761,455 x 168 "prime + prime" combinations
d) prime h + prime d: 5,761,455 x 1,229 "prime + prime" combinations
e) prime h + prime e: 5,761,455 x 9,592 "prime + prime" combinations
f) prime h + prime f: 5,761,455 x 78,498 "prime + prime" combinations
g) prime h + prime g: 5,761,455 x 664,579 "prime + prime" combinations
h) prime h + prime h: 5,761,455 x 5,761,455 "prime + prime" combinations
i) prime h + prime i: 5,761,455 x 50,847,534 "prime + prime" combinations
j) prime h + prime j: 5,761,455 x 455,052,512 "prime + prime" combinations

9) With (i) above, we have the following "prime + prime = even number" combinations:

a) prime i + prime a: 50,847,534 x 4 "prime + prime" combinations

b) prime i + prime b: 50,847,534 x 25 "prime + prime" combinations
c) prime i + prime c: 50,847,534 x 168 "prime + prime" combinations
d) prime i + prime d: 50,847,534 x 1,229 "prime + prime" combinations
e) prime i + prime e: 50,847,534 x 9,592 "prime + prime" combinations
f) prime i + prime f: 50,847,534 x 78,498 "prime + prime" combinations
g) prime i + prime g: 50,847,534 x 664,579 "prime + prime" combinations
h) prime i + prime h: 50,847,534 x 5,761,455 "prime + prime" combinations
i) prime i + prime i: 50,847,534 x 50,847,534 "prime + prime" combinations
j) prime i + prime j: 50,847,534 x 455,052,512 "prime + prime" combinations

10) With (j) above, we have the following "prime + prime = even number" combinations:

a) prime j + prime a: 455,052,512 x 4 "prime + prime" combinations
b) prime j + prime b: 455,052,512 x 25 "prime + prime" combinations
c) prime j + prime c: 455,052,512 x 168 "prime + prime" combinations
d) prime j + prime d: 455,052,512 x 1,229 "prime + prime" combinations
e) prime j + prime e: 455,052,512 x 9,592 "prime + prime" combinations
f) prime j + prime f: 455,052,512 x 78,498 "prime + prime" combinations
g) prime j + prime g: 455,052,512 x 664,579 "prime + prime" combinations
h) prime j + prime h: 455,052,512 x 5,761,455 "prime + prime" combinations
i) prime j + prime i: 455,052,512 x 50,847,534 "prime + prime" combinations
j) prime j + prime j: 455,052,512 x 455,052,512 "prime + prime" combinations

.
.
.
.

The above partitions/"prime + prime = even number" combinations are evidently progressively more "overwhelming", dense (refer to Figure 1 below), and repetitive (overlapping). *That is, the Goldbach conjecture becomes evidently progressively stronger and stronger towards infinity, which corroborates the earlier observation/induction.* It is not surprising that computer searches completed in 2000 had verified that all even numbers up to 400 trillion (4×10^{14}), which is not a small list, are sums of 2 primes, while in 2008, a distributed computer search ran by Tomas Oliveira e Silva, a researcher at the University of Aveiro, Portugal, had further verified the Goldbach conjecture up to 12×10^{17}, which is a long, impressive list.

Though the distribution of primes evidently becomes progressively less and less dense, e.g., ranging from 40% of primes within the first 10 integers to 4.55% of primes within the first 10,000,000,000 integers, the density of partitions/"prime + prime = even number" combinations evidently becomes progressively greater and greater as is shown below:-

1) For the 1st. **10-fold** increase in no. of integers (100 integers ÷ 10 integers), the no. of partitions/"prime + prime = even number" combinations increases **39.06 times** ([25 x 25 partitions] ÷ [4 x 4 partitions]).
2) For the 2nd. **10-fold** increase in no. of integers (1,000 integers ÷ 100 integers), the no. of partitions/"prime + prime = even number" combinations increases **45.16 times** ([168 x 168 partitions] ÷ [25 x 25 partitions]).
3) For the 3rd. **10-fold** increase in no. of integers (10,000 integers ÷ 1,000 integers), the no. of partitions/"prime + prime = even number" combinations increases **53.52 times** ([1,229 x 1,229 partitions] ÷ [168 x 168 partitions]).
4) For the 4th. **10-fold** increase in no. of integers (100,000 integers ÷ 10,000 integers), the no. of partitions/"prime + prime = even number" combinations increases **60.91 times** ([9,592 x 9,592 partitions] ÷ [1,229 x 1,229 partitions]).
5) For the 5th. **10-fold** increase in no. of integers (1,000,000 integers ÷ 100,000 integers), the no. of partitions/"prime + prime = even number" combinations increases **66.97 times** ([78,498 x 78,498 partitions] ÷ [9,592 x 9,592 partitions]).
6) For the 6th. **10-fold** increase in no. of integers (10,000,000 integers ÷ 1,000,000 integers), the no. of partitions/"prime + prime = even number" combinations increases **71.68 times** ([664,579 x 664,579 partitions] ÷ [78,498 x 78,498 partitions]).
7) For the 7th. **10-fold** increase in no. of integers (100,000,000 integers ÷ 10,000,000 integers), the no. of partitions/"prime + prime = even number" combinations increases **75.16 times** ([5,761,455 x 5,761,455 partitions] ÷ [664,579 x 664,579 partitions]).
8) For the 8th. **10-fold** increase in no. of integers (1,000,000,000 integers ÷ 100,000,000 integers), the no. of partitions/"prime + prime = even number" combinations increases **77.89 times** ([50,847,534 x 50,847,534 partitions] ÷ [5,761,455 x 5,761,455 partitions]).
9) For the 9th. **10-fold** increase in no. of integers (10,000,000,000 integers ÷ 1,000,000,000 integers), the no. of partitions/"prime + prime = even number" combinations increases **80.09 times** ([455,052,512 x 455,052,512 partitions] ÷ [50,847,534 x 50,847,534 partitions]).

<u>Figure 1</u>

The infinitude of the primes, as per Euclid's proof, together with the infinitude of the even numbers, however imply that the above partitions/"prime + prime = even number" combinations would become increasingly more "overwhelming", dense, and repetitive (overlapping) towards infinity (the Goldbach conjecture becoming evidently stronger and stronger the higher up the infinite list of prime numbers/even numbers we go), hence "ensuring" the continuity (without any breaks or gaps) of the even numbers generated, and would be so all the way to infinity, thus proving that every even number after 2 is the sum of 2 primes. (For a better insight of how the above partitions/"prime + prime = even number" combinations would become increasingly more "overwhelming", dense, and repetitive (overlapping) towards infinity, refer to Appendix 1 and Appendix 2.)

The partitions/"prime + prime = even number" combinations, as had been conjectured by Goldbach, are evidently effusive, or in great abundance, in their occurrences, as is shown above and in the appendices below. This has important consequence. For instance, in Appendix 2, the number of partitions/"prime + prime = even number" combinations for each of the 30 even numbers (424 to 482) ranges from the minimum 9 (for the even numbers 428 and 458) to the maximum 30 (for the even numbers 462 and 480), giving an average of 16.93 partitions/"prime + prime = even number" combinations per even number. This is significant and is in stark contrast to the results of the Fundamental Theorem of Arithmetic or Unique Factorization Theorem, which states that there is only 1 possible combination of primes which will multiply together to produce any particular number, e.g., the only combination of primes which will produce the number 2,079 is as follows:-

3 x 3 x 3 x 7 x 11 (only)

In the same manner, the following numbers are also uniquely factorized:-

63 = 3 x 3 x 7 (only)
153 = 3 x 3 x 17 (only)
1,021,020 = 2 x 2 x 3 x 5 x 7 x 11 x 13 x 17 (only)

In other words, every positive whole number can be broken up into prime factors, and, this can happen in only 1 way. In contrast, every even number is the sum of 2 primes in more than 1 way, e.g., 30 ways (i.e., 30 possible partitions) in the cases of the even numbers 462 and 480 as is described above. As is stated above, this is significant. This effusiveness or abundance of partitions/"prime + prime = even number" combinations somehow implies that the continuity (without any breaks or gaps) of the even numbers as sums of 2 primes (which are "generated" through the various additions of 2 primes) is "ensured", i.e., the possible breaks in the continuity of the even numbers as sums of 2 primes (wherein some even numbers in-between can never be sums of 2 primes, as are shown in the example in Figure 2 below where there are

4 breaks in the continuity of the even numbers x_1 to x_{12}, where the 4 even numbers x_4, x_5, x_8 & x_{10} can never be sums of 2 primes), which implies the falsity of the Goldbach conjecture, are somehow "prevented from happening" by this effusiveness or abundance:-

x below represents, say, an extremely large even number. p below represents a prime. c below represents a composite number or non-prime whence the Goldbach conjecture would be false (i.e., not every even number is the sum of 2 primes as the composite numbers or non-primes would be the exceptions).

$$
\begin{aligned}
&\;\;\;\;\;\;\;\;\;\;\bullet \\
&\;\;\;\;\;\;\;\;\;\;\bullet \\
&\;\;\;\;\;\;\;\;\;\;\bullet \\
x_1 &= p_1 + p_2 \\
x_2 &= p_3 + p_4 \\
x_3 &= p_5 + p_6 \\
x_4 &= c_1 + c_2 \,(\text{break}) \\
x_5 &= p_7 + c_3 \,(\text{break}) \\
x_6 &= p_8 + p_9 \\
x_7 &= p_{10} + p_{11} \\
x_8 &= c_4 + p_{12} \,(\text{break}) \\
x_9 &= p_{13} + p_{14} \\
x_{10} &= c_5 + c_6 \,(\text{break}) \\
x_{11} &= p_{15} + p_{16} \\
x_{12} &= p_{17} + p_{18} \\
&\;\;\;\;\;\;\;\;\;\;\bullet \\
&\;\;\;\;\;\;\;\;\;\;\bullet \\
&\;\;\;\;\;\;\;\;\;\;\bullet \\
&\;\;\;\;\;\;\;\;\;\;\bullet
\end{aligned}
$$

Figure 2

There appears to be some deep meaning in the ease and effusiveness with which the partitions/"prime + prime = even number" combinations or sums of 2 primes show up, as is shown in this chapter. If every even number which is the sum of 2 primes is the sum of 2 primes in only 1 way (a la the results of the Fundamental Theorem of Arithmetic or Unique Factorization Theorem described above), there could be possible breaks in the continuity of the even numbers as sums of 2 primes (as are shown in Figure 2 above), in other words, there could be some reason to doubt the validity of the Goldbach conjecture. But, on the contrary, the sums of 2 primes are evidently much effusive; they are evidently a defining characteristic of the even numbers. Under such a circumstance, it would be difficult to doubt the validity of the Goldbach conjecture.

In elaborating further on the above point, we take a look at the following:-

\No. Of Old/Repeated (Also Appeared Earlier) Even Numbers/Overlaps "Generated" (By The Additions/Combinations Of Two Primes), For Integers 1 To 1,250 (*See Appendix 1 For Example Of Computation Method*)

(a)   Set Of Integers, 1 To 50, With 14 Primes Within It = Not Applicable
(aa) Percentage Increase In Repetition = Not Applicable

(b)   Set Of Integers, 51 To 100, With 10 Primes Within It = **20** Repeated Even Nos.
(bb) Percentage Increase In Repetition = Not Applicable

(c)   Set Of Integers, 101 To 150, With 10 Primes Within It = 46 Repeated Even Nos.
(cc) Percentage Increase In Repetition = (46 - 20) ÷ 20 x 100% = **130%**

(d)   Set Of Integers, 151 To 200, With 11 Primes Within It = 73 Repeated Even Nos.
(dd) Percentage Increase In Repetition = (73 - 46) ÷ 46 x 100% = 58.7%

(e)   Set Of Integers, 201 To 250, With 7 Primes Within It = 93 Repeated Even Nos.
(ee) Percentage Increase In Repetition = (93 - 73) ÷ 73 x 100% = 27.4%

(f)   Set Of Integers, 251 To 300, With 9 Primes Within It = 115 Repeated Even Nos.
(ff)   Percentage Increase In Repetition = (115 - 93) ÷ 93 x 100% = 23.66%

(g)   Set Of Integers, 301 To 350, With 8 Primes Within It = 139 Repeated Even Nos.
(gg) Percentage Increase In Repetition = (139 - 115) ÷ 115 x 100% = 20.87%

(h)   Set Of Integers, 351 To 400, With 8 Primes Within It = 172 Repeated Even Nos.
(hh) Percentage Increase In Repetition = (172 - 139) ÷ 139 x 100% = 23.74%

(i)   Set Of Integers, 401 To 450, With 9 Primes Within It = 196 Repeated Even Nos.
(ii)   Percentage Increase In Repetition = (196 - 172) ÷ 172 x 100% = 13.95%

(j)   Set Of Integers, 451 To 500, With 8 Primes Within It = 220 Repeated Even Nos.

(jj)   Percentage Increase In Repetition = (220 - 196) ÷ 196 x 100% = 12.24%

(k)   Set Of Integers, 501 To 550, With 6 Primes Within It = 247 Repeated Even Nos.
(kk)   Percentage Increase In Repetition = (247 - 220) ÷ 220 x 100% = 12.27%

(l)   Set Of Integers, 551 To 600, With 8 Primes Within It = 268 Repeated Even Nos.
(ll)   Percentage Increase In Repetition = (268 - 247) ÷ 247 x 100% = 8.5%

(m)   Set Of Integers, 601 To 650, With 9 Primes Within It = 298 Repeated Even Nos.
(mm) Percentage Increase In Repetition = (298 - 268) ÷ 268 x 100% = 11.19%

(n)   Set Of Integers, 651 To 700, With 7 Primes Within It = 320 Repeated Even Nos.
(nn)   Percentage Increase In Repetition = (320 - 298) ÷ 298 x 100% = 7.38%

(o)   Set Of Integers, 701 To 750, With 7 Primes Within It = 340 Repeated Even Nos.
(oo)   Percentage Increase In Repetition = (340 - 320) ÷ 320 x 100% = 6.25%

(p)   Set Of Integers, 751 To 800, With 7 Primes Within It = 367 Repeated Even Nos.
(pp)   Percentage Increase In Repetition = (367 - 340) ÷ 340 x 100% = 7.94%

(q)   Set Of Integers, 801 To 850, With 7 Primes Within It = 392 Repeated Even Nos.
(qq)   Percentage Increase In Repetition = (392 - 367) ÷ 367 x 100% = 6.81%

(r)   Set Of Integers, 851 To 900, With 8 Primes Within It = 412 Repeated Even Nos.
(rr)   Percentage Increase In Repetition = (412 - 392) ÷ 392 x 100% = 5.1%

(s)   Set Of Integers, 901 To 950, With 7 Primes Within It = 433 Repeated Even Nos.
(ss)   Percentage Increase In Repetition = (433 - 412) ÷ 412 x 100% = 5.1%

(t)   Set Of Integers, 951 To 1,000, With 7 Primes Within It = 470 Repeated Even Nos.
(tt)   Percentage Increase In Repetition = (470 - 433) ÷ 433 x 100% = 8.55%

(u)   Set Of Integers, 1,001 To 1,050, With 8 Primes Within It = 492 Repeated Even Nos.
(uu)   Percentage Increase In Repetition = (492 - 470) ÷ 470 x 100% = 4.68%

(v)   Set Of Integers, 1,051 To 1,100, With 8 Primes Within It = 523 Repeated Even Nos.
(vv)   Percentage Increase In Repetition = (523 - 492) ÷ 492 x 100% = 6.3%

(w)   Set Of Integers, 1,101 To 1,150, With 5 Primes Within It = 545 Repeated Even Nos.
(ww) Percentage Increase In Repetition = (545 - 523) ÷ 523 x 100% = 4.21%

(x)   Set Of Integers, 1,151 To 1,200, With 7 Primes Within It = 553 Repeated Even Nos.
(xx)   Percentage Increase In Repetition = (553 - 545) ÷ 545 x 100% = **1.47%**

(y)   Set Of Integers, 1,201 To 1,250, With 8 Primes Within It = **592** Repeated Even Nos.
(yy)   Percentage Increase In Repetition = (592 - 553) ÷ 553 x 100% = **7.05%**

It could be seen above that on the whole the No. Of Old/Repeated (Also Appeared Earlier) Even Numbers/Overlaps "Generated" (By The Additions/Combinations Of Two Primes) increases progressively from 20 in (b) to 592 in (y), while it could be seen that the Percentage Increase In Repetition on the whole thins out from 130% in (cc) to 7.05% in (yy), with the lowest percentage increase of 1.47% found in (xx). This statistical trend or feature is not surprising and represents significant evidence that lends support to the validity of the Goldbach conjecture - the infinitude of both the primes and the even numbers implies that the above overlaps increase progressively to infinity.

It is evident here that the higher up the primes we go the more "overwhelmingly" the even numbers "generated" would repeat themselves and overlap. This is significant. Though the infinitude of the prime numbers would ensure that there would always be new even numbers being "generated", there is also the "fear" that there might be gaps, breaks or lack of continuity in the even numbers thus "generated" wherein some of the even numbers in-between can never be sums of 2 primes (as are shown in the example in Figure 2 above), thereby disproving the Goldbach conjecture. But, it is evident that these more and more profuse repetitions and overlaps of the even numbers thus "generated" by the primes the higher up the infinite list of prime numbers we go "ensure" that such gaps or breaks would not appear between the even numbers "generated" - they "ensure" that the even numbers thus "generated" by the primes in the infinite list of primes would be regular, continuous, *without breaks or gaps*, and, in consecutive running order. This evident greater and greater effusiveness or exuberance of the repetitions and overlaps of the even numbers thus "generated" by the primes the higher up the infinite list of prime numbers we go can be likened to a "play-safe measure" wherein there is "safety derived from large numbers". In other words, since an even number could be formed in so many ways by adding 2 primes, i.e., so easily formed by adding 2 primes, evidently more so the higher up the infinite list of prime numbers we go, as has been shown above, the sums of 2 primes thus becoming evidently a defining characteristic of the even numbers, we could expect every larger and larger even number to be the sum of 2 primes in more and more ways as has been shown above (and in the appendices below).

We note again that a long, impressive list of consecutive even numbers, from 4 to 12×10^{17}, had already been verified to be sums of 2 primes, and, these partitions/"prime + prime = even number" combinations would become increasingly more "overwhelming", dense, and repetitive (overlapping) towards infinity (the Goldbach conjecture becoming evidently stronger and stronger the higher up the infinite list of prime numbers/even numbers we go), as is described above. The moot question now is, of course, whether after 12×10^{17} there would be an even number in the infinite list of even numbers which is the last, or, largest, even number that is the sum of 2 primes - this largest even number, if it exists (thereby proving the falsehood of the Goldbach conjecture), must (of necessity) be the sum of 2 primes that are each the largest existing prime. However, as the primes are infinite (vide Euclid's proof), a largest existing prime is an impossibility. Therefore, there can never be a largest even number comprising of the summation of 2 largest existing primes which would disprove the Goldbach conjecture. As a matter of fact, the infinity of the primes implies that there would be an infinite number of double primes which sum up to an even number.

The Goldbach conjecture is thus valid.

CONCLUDING REMARKS

A number of methods have been adopted in this chapter in proving the Goldbach conjecture.

The inductive method, which is a well-established proof, is one of the methods utilized. The following lends support to this inductive proof of the Goldbach conjecture: (a) The characteristic of a mountain or infinite volume of sand is reflected in the characteristic of some grains of sand found there so that studying the characteristic of some grains of sand found there is enough for deducing the characteristic of the mountain or infinite volume of sand, to ascertain the quality of a batch of products it is only necessary to inspect some carefully selected samples from that batch of products and not every one of the products and to carry out a population census, i.e., find out the characteristics of a population, it is only necessary to carry out a survey on some carefully selected respondents and not the whole population; in like manner, by the same principle, we just need to study a carefully selected list of even numbers, find out whether they are all sums of 2 primes and deduce by induction whether all even numbers after this list would also be sums of 2 primes - this act is rather like extrapolation. (For example, a distributed computer search completed in 2008 at the University of Aveiro, Portugal, had confirmed that every even number up to 12×10^{17}, which is no small list of numbers, is the sum of 2 primes. By the principle of induction in this case we could deduce that all the even numbers after 12×10^{17} would also be sums of 2 primes.) (b) Thus, in this way every even number after 2 could be reasonably proved to be the sum of 2 primes. In fact, induction plays an important part in the proof.

The other argument used to prove the conjecture is the indirect (reductio ad absurdum) method, which had been used by Euclid and other mathematicians after him. Logically, 1 or 2 examples of "contradiction" should be sufficient proof of infinity, for it does not make sense to have a need for an infinite number of cases of "contradiction", as our proof would then have to be infinitely and impossibly long, an absurdity. This method of proof is "proof by implication" as a result of "contradiction" - which is a "short-cut" and smart way in proving infinity, instead of "proving infinity by counting to infinity", which is ludicrous, and, impossible. Hence, 1 or 2 cases of "contradiction" should be sufficient for implying that there would be an infinitude of even numbers which are sums of 2 primes, which of course also tacitly implies that there would be an infinitude of the number of cases of such "contradiction". (Euclid evidently had this logical point in mind when he formulated the indirect (reductio ad absurdum) proof of the infinity of the primes.) This method of proof had been cleverly used by a number of mathematicians, not the least by the great German mathematician, David Hilbert. For example, Hilbert had used an indirect method (the "reductio ad absurdum" proof) to prove Gordan's Theorem without having to show an actual "construction", a proof which had been accepted by his peers.

One important query here, which many might not have considered: What if the list of prime numbers is not infinite? Of course, if that is the case, the Goldbach conjecture would be false. It would then have been absurd for the Goldbach conjecture to have been conceived at all. However, the list of primes is infinite (vide Euclid's proof). This gives credence to the Goldbach conjecture.

A very important related point must be highlighted here. If the Goldbach conjecture were indeed false, there must be an ultimate (largest) even number which is (and must necessarily be) the result of the summation of 2 primes that are each the largest existing prime. It must be noted that this is actually an impossibility, as there can never be a largest existing prime - by Euclid's proof, the primes are infinite (refer to argument just above). Hence, the Goldbach conjecture cannot be false, and, by both reductio ad absurdum (contradiction), and, induction (wherein all even numbers up to 12×10^{17}, a long, impressive list, had been confirmed to be sums of 2 primes), has to be true.

Another very important point is that the Goldbach conjecture becomes evidently stronger and stronger the higher up the infinite list of prime numbers/even numbers we go, as has been shown above. Thus, by implication, induction, extrapolation, it could be concluded that the Goldbach conjecture is valid - that every even number after 2 is the sum of 2 primes.

So far, there has been no indication or confirmation at all that the number of even numbers after the number 2 which are each the sum of 2 primes is finite and the largest existing even number which is the sum of 2 primes has not been found and confirmed. (This would of course be the case if the

Goldbach conjecture is true.) Also, no counter-example (i.e., an even number which is never the sum of 2 primes) has been found so far. On the other hand, practically everyone could intuit that the list of even numbers after the number 2 which are each the sum of 2 primes is infinite. Besides, the evidence, as shown in this chapter, is strongly in support of the infinity of this list.

APPENDIX 1

(20) Set Of Integers, 1,201 To 1,250, With 8 Primes Within It
    (a) Primes: 1,201; 1,213; 1,217; 1,223; 1,229; 1,231; 1,237 and 1,249
    (b) No. Of Primes: 8
    (c) No. Of Even Numbers "Generated" (Including Repetitions) By The 8 Primes = 648 (1,204 [1,201 + 3] To 2,498 [1,249 + 1,249])
    (d) No. Of New Even Numbers "Generated" = 56 (2,388 To 2,498)
    (e) No. Of Old/Repeated (Also Appeared In (19) Above, With Some Also Having Appeared In (18), (17), (16), (15), (14), (13), (12), (11), (10), (9) And (8) Above) Even Numbers "Generated" (I.e., Repetitions/Overlaps) = 592 (1,204 To 2,386)
    (f) Density Of New Even Numbers "Generated" = (d) ÷ 8 Primes = 56 ÷ 8 = 7 New Even Numbers Per Prime Number

APPENDIX 2

(8) 8 th. Batch Of 30 Even Numbers (424 To 482) - Partitions/"Prime + Prime = Even Number" Combinations
    (a) 424: No. Of Above-mentioned Prime Additions/Combinations = 12
    (b) 426: No. Of Above-mentioned Prime Additions/Combinations = 21
    (c) 428: No. Of Above-mentioned Prime Additions/Combinations = **9**
    (d) 430: No. Of Above-mentioned Prime Additions/Combinations = 14
    (e) 432: No. Of Above-mentioned Prime Additions/Combinations = 19
    (f) 434: No. Of Above-mentioned Prime Additions/Combinations = 14
    (g) 436: No. Of Above-mentioned Prime Additions/Combinations = 11
    (h) 438: No. Of Above-mentioned Prime Additions/Combinations = 22
    (i) 440: No. Of Above-mentioned Prime Additions/Combinations = 15
    (j) 442: No. Of Above-mentioned Prime Additions/Combinations = 13
    (k) 444: No. Of Above-mentioned Prime Additions/Combinations = 22
    (l) 446: No. Of Above-mentioned Prime Additions/Combinations = 12
    (m) 448: No. Of Above-mentioned Prime Additions/Combinations = 13
    (n) 450: No. Of Above-mentioned Prime Additions/Combinations = 29

(o) 452: No. Of Above-mentioned Prime Additions/Combinations = 14
(p) 454: No. Of Above-mentioned Prime Additions/Combinations = 12
(q) 456: No. Of Above-mentioned Prime Additions/Combinations = 26
(r)   458: No. Of Above-mentioned Prime Additions/Combinations = **9**
(s) 460: No. Of Above-mentioned Prime Additions/Combinations = 17
(t) 462: No. Of Above-mentioned Prime Additions/Combinations = **30**
(u) 464: No. Of Above-mentioned Prime Additions/Combinations = 13
(v) 466: No. Of Above-mentioned Prime Additions/Combinations = 14
(w) 468: No. Of Above-mentioned Prime Additions/Combinations = 26
(x) 470: No. Of Above-mentioned Prime Additions/Combinations = 16
(y) 472: No. Of Above-mentioned Prime Additions/Combinations = 14
(z) 474: No. Of Above-mentioned Prime Additions/Combinations = 24
(aa) 476: No. Of Above-mentioned Prime Additions/Combinations = 14
(bb) 478: No. Of Above-mentioned Prime Additions/Combinations = 12
(cc) 480: No. Of Above-mentioned Prime Additions/Combinations = **30**
(dd) 482: No. Of Above-mentioned Prime Additions/Combinations = 11
(i)   Maximum No. Of Prime Additions/Combinations = 30
(ii) Minimum No. Of Prime Additions/Combinations = 9
(iii) Total No. Of Prime Additions/Combinations For (a) To (dd) = 508
(iv) Total No. Of Even Numbers = 30
(v) Density Of Distribution = Average Prime Additions/Combinations Per Even Number = (iii) ÷ (iv) = 508 ÷ 30 = 16.93 Prime Additions/Combinations Per Even Number

15 A PROOF OF POLIGNAC'S CONJECTURE

There are infinitely many cases of two consecutive prime numbers with difference n [n = 2, 4, 6, 8, ….].

Lemma: A fraction of infinity is also infinite. (See appendix.)

The list of the primes, which are the building-blocks of the integers, had been proven by Euclid to be infinite.

It would be appropriate to conduct an examination of the primes pairs separated by 2 integers, 4 integers, 6 integers, 8 integers and more. For this purpose, we select a reasonably large list of consecutive primes, which may be regarded as a basic unit of the infinite list of the primes; we examine a compilation of such data obtained from, say, the list of the first 2,500 consecutive primes - 2 to 22,307 - which is as follows:-

LIST OF PRIMES PAIRS FOR THE FIRST 2,500 CONSECUTIVE PRIMES, 2 TO 22,307, RANKED ACCORDING TO THEIR FREQUENCIES OF APPEARANCE

S. No.	Ranking	Prime Pairs	No. Of Pairs	Percentage
(1)	1	primes pair separated by 6 integers	482	19.29 %
(2)	2	primes pair separated by 4 integers	378	15.13 %
(3)	3	primes pair separated by 2 integers (t. p.)	376	15.05 %
(4)	4	primes pair separated by 12 integers	267	10.68 %
(5)	5	primes pair separated by 10 integers	255	10.20 %
(6)	6	primes pair separated by 8 integers	229	9.16 %
(7)	7	primes pair separated by 14 integers	138	5.52 %
(8)	8	primes pair separated by 18 integers	111	4.44 %
(9)	9	primes pair separated by 16 integers	80	3.20 %
(10)	10	primes pair separated by 20 integers	47	1.88 %
(11)	11	primes pair separated by 22 integers	46	1.84 %
(12)	12	primes pair separated by 30 integers	24	0.96 %
(13)	13	primes pair separated by 28 integers	19	0.76 %
(14)	14	primes pair separated by 24 integers	16	0.64 %
(15)	15	primes pair separated by 26 integers	10	0.40 %
(16)	16	primes pair separated by 34 integers	9	0.36 %
(17)	17	primes pair separated by 36 integers	5	0.20 %
(18)	18	primes pair separated by 32 integers	2	0.08 %
(19)	18	primes pair separated by 40 integers	2	0.08 %
(20)	19	primes pair separated by 42 integers	1	0.04 %
(21)	19	primes pair separated by 52 integers	1	0.04 %

Total No. Of Primes Pairs In List: 2,498

It is evident in the above list that the primes pairs separated by 6 integers, 4 integers and 2 integers (twin primes), among the 21 classifications of primes pairs separated by from 2 integers to 52 integers (primes pairs separated by 38 integers, 44 integers, 46 integers, 48 integers & 50 integers are not among them, but, they are expected to appear further down in the infinite list of the primes), are the most dominant, important. There is a long list of other primes pairs, besides those shown in the above list, which also play a part as the building-blocks of the infinite list of the integers. We shall prove the infinity of all these various building-blocks below.

The list of the integers is infinite. The list of the primes is also infinite. The infinite primes are the building-blocks of the infinite integers - the infinite odd integers are all either primes or composites of primes, and, the infinite even integers, except for 2 which is a prime, are all also composites of primes. Therefore, all the primes pairs separated by the integers of various magnitudes, as described above, could never all be finite.

All of the above lists of primes pairs which are respectively separated by from 2 integers to 52 integers are each respectively a fraction of the infinite list of the primes: the list of primes pairs separated by 2 integers (twin primes) is a fraction of the infinite list of the primes, the list of primes pairs separated by 4 integers is also a fraction of the infinite list of the primes, the list of primes pairs separated by 6 integers is a fraction of the infinite list of the primes too, and so on all the way down to the list of primes pairs separated by 52 integers, et al.

Therefore, by the above lemma, all these various lists of primes pairs separated by integers of various magnitudes are each also infinite.

APPENDIX - EUCLID'S PROOF OF THE INFINITY OF THE PYTHAGOREAN TRIPLES: A FRACTION OF INFINITY IS ALSO INFINITE

A Pythagorean triple is a set of 3 integers wherein 1 number squared added to another number squared equals the third number squared, e.g., 3^2 (9) + 4^2 (16) = 5^2 (25) below.

Euclid's proof of the infinity of the Pythagorean triples begins with the statement that the difference between 2 successive square numbers is always an odd number, as is evident below:

Column 1	Column 2	Column 3
1^2	1	
		} 3 (difference between 4 & 1)
2^2	4	
		} 5 (difference between 9 & 4)
3^2	9	
		} 7 (difference between 16 & 9)
4^2	16	
		} 9 (difference between 25 & 16)
5^2	25	
		} 11 (difference between 36 & 25)
6^2	36	
		} 13 (difference between 49 & 36)
7^2	49	
		} 15 (difference between 64 & 49)
8^2	64	
		} 17 (difference between 81 & 64)
9^2	81	
		} 19 (difference between 100 & 81)
10^2	100	
.	.	.
.	.	.
.	.	.

Every one of the infinity of odd numbers (in Column 3 above) could be added to a particular square number (in Column 2 above) to give another square number, e.g., 3 in Column 3 above could be added to 1 in Column 2 above to give the square number 4, 5 in Column 3 above could be added to 4 in Column 2 above to give the square number 9, 7 in Column 3 above could be added to 9 in Column 2 above to give the square number 16, and so on .... A fraction of the infinite odd numbers in Column 3 above are themselves square numbers, e.g., the odd number 9 in Column 3 above is a square number, and, is the only square number in the list of odd numbers shown there, representing there a fraction of the infinite odd numbers which are square numbers. However, a fraction of infinity is also infinite.

Therefore, there are also an infinity of odd square numbers (in Column 3 above) which could each be added to another square number (in Column 2 above) to give another square number, i.e., there is an infinitude of Pythagorean triples.

16 THE NAVIER-STOKES EQUATIONS

The Navier-Stokes differential equations describe the motion of fluids which are incompressible. The three-dimensional Navier-Stokes equations misbehave very badly although they are relatively simple-looking. The solutions could wind up being extremely unstable even with nice, smooth, reasonably harmless initial conditions. A mathematical understanding of the outrageous behaviour of these equations would dramatically alter the field of fluid mechanics. This chapter describes why the three-dimensional Navier-Stokes equations are not solvable, i.e., the equations cannot be used to model turbulence, which is a three-dimensional phenomenon.

The general equations of motion for a viscous fluid were obtained by Sir George Stokes in 1845. The following is the fundamental equation (in vectorial form) governing the flow of a viscous fluid:-

$$\partial v / \partial t + (v . \bigtriangledown) v = - 1/p \bigtriangledown Pe - \bigtriangledown \varphi + \eta/p \bigtriangledown 2v \, ,$$

where v is the velocity of the fluid (as a function of position), Pe the pressure, φ the gravitational potential, p the density and η the viscosity.

The scientist normally makes a forecast of the outcome of a flow and uses the Navier-Stokes equations to model this forecast. However, in the instance of turbulence, making this forecast will be fraught with difficulty, if it can be carried out at all. Putting it another way, if turbulence could be forecasted, predicted and described by the Navier-Stokes equations it could not be turbulence, for turbulence implies puzzlement, lack of order or pattern and lack of predictability.

The Navier-Stokes equations are nonlinear due to the acceleration terms such as $u \partial u / \partial x$. As a result, the solution to these equations may not be unique. For instance, the flow between two rotating cylinders can be solved using the Navier-Stokes equations to treat a relatively simple flow with circular streamlines; it can also be a flow with streamlines which are like a spring wound around the cylinders as a torus; there are also more complex flows which are solutions to the Navier-Stokes equations, all satisfying the identical boundary conditions.

For simple geometries, the Navier-Stokes equations can be solved with relative ease. However, the equations cannot be solved for a turbulent flow even for the simplest of examples. A turbulent flow is highly unsteady, nonlinear and three-dimensional and therefore requires that the three velocity components be specified at all points in a region of interest at some initial time, say $t = 0$. But, even for the simplest geometry, such information will be almost impossible to obtain. Therefore, the solutions for turbulent flows have to be left to the experimentalist and are not attempted by solving the Navier-Stokes equations.

17 THEORY OF TURBULENCE

The motion of fluids which are incompressible could be described by the Navier-Stokes differential equations. However, the three-dimensional Navier-Stokes equations for modelling turbulence misbehave very badly although they are relatively simple-looking. The solutions could wind up being extremely unstable even with nice, smooth, reasonably harmless initial conditions. A mathematical understanding of the outrageous behaviour of these equations would greatly affect the field of fluid mechanics. In this chapter, a reasoned, practical approach towards resolving the issue is adopted and a practical, statistical kind of mathematical solution is proposed.

Navier-Stokes Equation
The general equations of motion for a viscous fluid had been obtained by Sir George Stokes in 1845. The fundamental equation (in vectorial form) governing the flow of a viscous fluid is as follows:-

$$\partial v/\partial t + (v.\nabla)v = -1/p\nabla Pe - \nabla\varphi + \eta/p\nabla 2v,$$

where v is the velocity of the fluid (as a function of position), Pe the pressure, φ the gravitational potential, p the density and η the viscosity.

A fluid in motion could be characterised by its velocity field (velocity as a function of position). However, because of the complex nature of the forces affecting fluids (in general, forces of both compression and viscosity) the result of applying basic principles such as Newton's second law is a set of nonlinear equations. Computational methods therefore play a large part in fluid dynamics. (Newton's second law states that the rate of change of momentum p of a body equals the total force F acting upon it, as is described by the following equation:

$$F = \partial p/\partial t.$$

If, as is normally the case, the mass of the body is constant, $F = \partial(mv)/\partial t$ reduces to $F = m\partial v/\partial t$ or $F = ma$, where a is the acceleration of the body. Note that the force and acceleration are vectors. The first law is the null case of the second law (if $F = 0$ then $a = 0$).)

The Navier-Stokes equation is a miracle of brevity, relating a fluid's velocity, pressure, density and viscosity. In two dimensions, fluid flow governed by this partial differential equation is deterministic and predictable. But this equation fails when the fluid becomes turbulent as turbulence represents three-dimensional flow of the fluid, for which the Navier-Stokes equation does very poorly. Whereas fluid flow under normal conditions tends to be laminar, in turbulence it becomes irregular and develops eddies, ripples and whorls. But yet there is some sort of order found within this disorder or turbulence which could be described as self-similar or fractal. What mathematical technique could be used to describe this state?

The Navier-Stokes equations are nonlinear and do not submit to any general method of solution. Each new problem has to be carefully formulated as to geometry and proper boundary conditions. Then some scheme of attack might be adopted with the hope of reaching a solution. In most cases all attempts to obtain an exact solution fail. Approximate solutions have to make do. In a few cases exact solutions could be obtained. The possibility that perhaps the flow of the fluid is unidirectional, i.e., $v(x, y, t) = 0$, is not an assumption. It is rather an intuitive guess which is pursued until we either find a solution or become convinced that it does not lead to a solution, in which case we mark it as an unsuccessful trial.

Substitution of viscosity in the Navier-Stokes equations with viscosity $= 0$ reduces them to a form called the Euler equations:

$$p \, Dq/Dt = pg - \nabla p \quad \text{(in vectorial form)}$$

The Euler equations had been formulated earlier than the Navier-Stokes equations and considered an approximation. The Euler equations are of the first order and cannot in general satisfy the boundary conditions. We could therefore conclude that the Euler equations do not form a good approximation near a rigid boundary. Far from a boundary and where viscosity $= 0$ is a fair estimate, they have an important role as approximations and are generally easier to solve than the full Navier-Stokes equations.

The Navier-Stokes equations do need for their solution initial conditions as well as boundary conditions. The following are proper boundary conditions for a velocity on a rigid boundary:

$$q_n = q_t = 0 \, ,$$

where q_n is the normal component of the velocity relative to the solid boundary, and q_t is the tangential component. These conditions are also termed the no-penetration ($q_n = 0$) and no-slip ($q_t = 0$) viscous boundary conditions. When the region occupied by the fluid is not closed, i.e., the fluid is not completely confined, additional conditions are still required on some surfaces which completely enclose the domain of the solution. These might represent some real physical surfaces or they might be chosen quite arbitrarily, provided the velocity on them is known. The pressure, which is also a dependent variable, also requires boundary conditions. The Navier-Stokes equations are then satisfied and we now know the resulting pressure field. This flow can exist only if the obtained pressure is possible. An acceptable boundary condition might be:

$$p = p_\infty = \text{const at } r \to \infty,$$

which then implies: $p = p_\infty - pQ^2/8\Pi^2 \cdot 1/r^2$.

We also note that in the solution for the pressure there is no trace of the viscosity. This pressure therefore also satisfies the Euler equations. (As viscosity in a fluid enables it to smooth out or overcome the ripples, eddies and whorls of turbulence, a viscous fluid is in effect not so much affected by turbulence than a non-viscous fluid. Thus, the Navier-Stokes equations, as they relate to viscous fluids, present a better solution for incompressible fluids which are viscous and subject to turbulence than the Euler equations for non-viscous fluids.)

Newton's Law Of Viscosity
The following equation is known as Newton's law of viscosity:-

$$T_{yx} = \mu \, du/dy$$

A fluid that obeys this law is called a Newtonian fluid. This equation states that in unidirectional flow the shear stress in a Newtonian fluid is directly proportional to the transverse velocity gradient, du/dy, which is also known as the rate of shear strain or the rate of shear deformation.

There is no obvious reason why real fluids should obey this law. In fact, there are more fluids that do not obey this equation than those that do. Fluids that do not obey this law are called non-Newtonian. It is fortunate that the three most abundant fluids, air, water and petroleum, obey Newton's law of viscosity rather closely. The typical non-Newtonian fluids are paints, polymer solutions and melts, blood and many liquid food products, such as jellies, soups, etc.

Computing Of Reynolds Numbers
The following is the formula for the Reynolds number:-

$$Re = v \, L \, p \, / \, n,$$

where p (kg/m^3) is the fluid's density, v (m/s) is a typical fluid velocity, L (m) is some characteristic length, e.g., diameter of a pipe through which the fluid flows, and, n (kgm/s^2) is the coefficient of viscosity (Gooey substances have a higher n value than runny ones like petrol. Viscous force = ability to smooth out turbulent whorls, eddies and ripples.).

The Reynolds number is the ratio between two forces, the initial force and the viscous force (frictional force). When the Reynolds number is below a few hundred, the flow of the fluid is smooth. When the Reynolds number exceeds about 2000 to 3000, the flow is completely turbulent. Between these values, the flow is sometimes smooth, sometimes turbulent. The greater the viscosity of the fluid is, the greater is its capability in overcoming or smoothing out the whorls, eddies and ripples of turbulence.

Fluid Flow
According to the Law of Continuity, water flowing from a wide pipe to a narrower pipe speeds up, while water flowing from a narrow pipe to a wider one slows down, and, the slow-moving water in the wide pipe would always have a higher pressure than the fast-moving water in the narrower pipe. Turbulent fluid flow could cause a pipe to give way.

Some equations pertaining to fluid flow are as follows:-

(i) Energy (E) = Mass (m) x Speed (v^2)

(ii) Altitude (A) + Energy (E) = Constant

(iii) Energy (E) = Density of Fluid (p) x Speed (v^2)

(iv) Pressure (P) + Energy (E) = Constant

(v) Pressure (P) + Density of Fluid (p) x Speed (v^2) = Constant

Model For Turbulence
The logistic equation has been a popular method used for the modelling of turbulence and chaos. The formula had been invented by the Belgian mathematician, Pierre Francois Verhulst, and is very simple. But because the process has to be repeated over and over again it ends up being extremely complicated. The equation works in the following way:-

For instance, if X is the population now, then the population next year is given by:

$$Xnext = rX (1 - X),$$

where r is some constant which could be adjusted according to the population being modelled. It is simplest if values of X between 0 and 1 are taken, so that 1 is the maximum population and 0 represents extinction. We might, for instance, take an arbitrary value for r of 2.6, and begin thus.

Let X = 0.2. Then 1 - X = 0.8, and, X (1 - X) = 0.2 x 0.8 = 0.16.

Then multiply this result by 2.6 and we would get 0.416.

Repeat the process. Start with X = 0.416 and we would get 0.6317. The population increases.

Start with 0.6317 and we would get 0.6049. The population falls.

Start with 0.6049 and we would get 0.6214. The population goes up again.

Repeating or iterating this process over and over again we would obtain the following population figures:

0.6117, 0.6176, 0.6141, 0.6162, 0.6150, 0.6156, 0.6152, 0.6155, 0.6153, 0.6154, 0.6153, 0.6154, 0.6154, 0.6154.

The population rises and falls but converges on a fixed number.

Scientists have however tried to model turbulence or chaos using this equation. But, does this model really describe turbulence or chaos?

Analysing Motions Of Fluids With Fourier Analysis
The surface water in a wave moves in a circular path at an angular velocity $w = 2\pi/T$ where T is the period of rotation. Deeper water moves in ellipses of decreasing size and increasing eccentricity.

The superpositions of simple sinusoidal oscillations in fluids could produce more complicated patterns of oscillations. The inverse mathematical operation of Fourier analysis could reduce any complicated oscillation into a sum of its simple sinusoidal components, each with a different period and amplitude.

With waves, there is a phenomenon which oscillates both in time and space. It might seem that this would considerably complicate any mathematical attempt to describe the superposition of waves. We could practically analyse complicated wave shapes either by freezing them in time or by freezing them in space. In the time domain, we obtain the wave's frequency components, while in the space domain we get the corresponding spectrum of wavelengths. These two approaches could each stand on their own, one being transformable to the other, because the product fλ is a constant (the wave speed), the longer wavelengths corresponding to the lower frequencies (i.e., the longer periods). The wavelength spectrum could be computed directly from the frequency spectrum by noting that for each harmonic component $\lambda = v/f = v/T$ - this holds for all of the wave's Fourier components (e.g., $\lambda_o = 3$ metres, $f_0 = 20$ cycles per second and wave speed $v = 60$ m/s).

Fourier Series And Circles

We could interpret a Fourier series geometrically as the projection of a system of superimposed circular motions. A circle rides upon the nest of spinning circles beneath it, and, we project the motion of a point on its circumference onto a line. The result is a periodic but distinctly non-sinusoidal motion which could be described mathematically as follows:-

$$y(t) = (1) \sin(2\pi t/T) + (1/3) \sin(3.2\pi t/T) + (1/5) \sin(5.2\pi t/T) + (1/7) \sin(7.2\pi t/T)$$

where T = fundamental period of oscillation = time for one rotation of the largest circle

This is the Fourier series which has been truncated after the fourth term for greater simplicity. If more terms are added (i.e., more circles turning upon circles), the resulting graph would more nearly approximate a series of alternating horizontal line segments.

Probability Waves And Turbulence

We now examine several related important ideas in quantum theory. Schrodinger had found an equation that could be applied to any physical system in which the mathematical form of the energy is known, which is as follows:-

$$(\partial^2 \Psi/\partial x^2) + (8\pi^2 m/h^2)(E - V)\psi = 0$$

where ∂^2 is the second derivative with respect to x, x is the position of the particle, ψ is the Schrodinger wave function, or, the probability amplitude for an electron in the state n to scatter into the direction m, E is energy and V is potential energy.

The Schrodinger equation is a deterministic time-symmetrical description of nature. In classical mechanics, when one says that a quantum system is in a particular "state", one means that the state is a point in phase space. It is here described by a wave function whose evolution over time is expressed by the following equation:-

$$ih / 2\pi \, \partial \psi(t) / \partial t = H_{op} \psi(t)$$

This equation identifies the time derivative of the Schrodinger wave function ψ with the action of the Hamiltonian operator on ψ. It is not derived but assumed at the start, and could thus be validated only by experiment. In quantum theory, it is the fundamental law of nature. Here, ψ is the probability amplitude for an electron - it is only an abstraction and has no physical reality. ψ is also, in a sense, the electron's own intensity wave. When it is squared and the absolute value is taken, it turns out to be a physical probability of the associated particle's presence.

Later, Born stated that the probability of the existence of a state is given by the square of the normalised amplitude of the individual wave function (i.e., ψ^2). This was another new concept, i.e., the probability that a certain quantum state exists. Born had said there were no more exact answers in atomic theory, but just probabilities. The wave Ψ determines the likelihood that the electron would be in a particular position, and has no physical reality unlike the electromagnetic field.

Dirac had posited that light could be treated as waves or particles. In fact, in quantum mechanics, particles are regarded as waves. The behaviour of these particles could be predicted, as it were, and, they are therefore known as probability waves or Dirac wave particles. A wave/particle duality is present. However, when the particle is not observed, it remains a wave (a probability wave), but upon being observed it becomes a particle.

The following is the formal solution of the Schrodinger equation:-

$$\psi(t) = U(t)\,\psi(0)$$

where $U(t) = e^{-iHt}$, $U(t)$ is the evolution operator which links the value of the wave function at time t to that at the initial time t = 0. Both future and past play the same part, since $U(t_1)\,U(t_2) = U(t_1 + t_2)$, whatever the sign of t_1 and t_2. This property defines a dynamical group.

This description of how quantum particles behave could not be strictly applied to the macro- world of fluid flow. Despite this, the above "probability" principles pertaining to how quantum particles behave could be somewhat broadly adapted and used as a guide for the interpretation of turbulent fluid flow.

In another important theory, the Uncertainty Principle, which was propounded by Heisenberg, it is posited that the very act of observing a quantum particle affects its behaviour. According to this theory, the position and the momentum of an elementary particle could not be known simultaneously. The reason for this is that if an electron could be held still long enough for its position to be determined, then its momentum could no longer be determined. A special point is that the product of two uncertainties (or spreads of possible values) is always at least a certain minimum number. From the de Broglie/Einstein relation, $\Delta p \sim h/\lambda$, Heisenberg obtained the imprecision in the momentum. Multiplying the two inaccuracies together, he showed that the product, $\Delta x \, \Delta p$, would always be greater than or equal to ($\geq$) a certain amount, which is as follows:-

(i) $(\Delta x)(\Delta p) \geq (\lambda)(h/\lambda) \geq h$, or, …..

(ii) $\Delta x \, \Delta p \geq h$

where Δp and h/λ represent the de Broglie relation, and, Δx and λ are from the diffraction limit.

The frustrated researcher looking for certainty must always make a compromise, knowledge gained about time, for instance, is paid for in uncertainty about frequency and vice versa. Though one does not notice Heisenberg Uncertainty Principle in one's everyday experience with the gross macroscopic world, the wave/particle duality defeats the atomic experimentalist who looks for perfection.

Another important implication of uncertainty which is worthy of comment is its effect on causality - the relation of cause to effect. Cause produces an effect. In classical physics if one understands fully the nature of a particular cause, one could then predict the effect. Cause and effect and predictability were cornerstones of classical physics and now they were under question. If it is impossible to measure precisely both the position and velocity of an electron (or any other particle) at the same moment, then it is also impossible to predict exactly where that electron would be at any given time afterward. An experimenter could send off two electrons in the same direction, at the same speed, and they would not necessarily end up in the same location. In the language of physics, the same cause could bring about different effects. There are serious philosophical consequences in this idea.

All this in effect represents "chaotic" behaviour in the quantum world. We have to remember that chaotic behaviour is beyond prediction; the above-mentioned describes the behaviour of quantum particles in a probabilistic, uncertain sort of way - it represents a branch of physics known as statistical mechanics. The corollary of this is the macro-world phenomenon of three-dimensional turbulent fluid flow which defies the solution of the Navier-Stokes equations. Perhaps, a probability function, Φ, should be incorporated in the Navier-Stokes equation, like the case for the quantum particles.

Nature Of Turbulence
It has been found that the transition from steady state through several splittings or bifurcations to chaos is similar to phase transitions - transitions that take place when a substance changes from a gas to a liquid or a liquid to a solid - all these transitions are also similar in that scaling is involved. It is believed that the presence of strange attractors, which could be defined as an endless path in phase space where the future

depends sensitively on the initial conditions, is responsible for the presence of turbulence or chaos. A strange attractor possesses the following characteristics:-

i)   It is generated by a simple set of differential equations.
ii)   It attracts and therefore all nearby trajectories in phase space converge toward it.
iii)   It has a great or very sensitive dependence on initial conditions, i.e., tiny differences or errors in the initial conditions lead quickly to large differences in the trajectory.
iv)   It is fractal, i.e., there is self-similarity or some familiar pattern within it.

The problem with fluid flow in conditions of turbulence is that the path taken by the fluid is continuous but nowhere differentiable. Turbulence gives rise to whorls, eddies and ripples in the fluid. However, there is a self-similar structure or pattern within the fluid - whorls would be found within whorls. This is in accordance with the well-established self-similarity concept which had been developed by Mitchell Feigenbaum in the 1970s and which brought him fame. According to this concept, there is a tendency of identical mathematical structures to recur on many levels, and, within a given structure there would be smaller copies of the same structure, their sizes being determined by the scaling factor, which is 4.669 and found to be a constant like pi (3.142). (This means there is some kind of order or pattern found in turbulence, chaos or disorder.) If we were to plot a curve to describe the fluid's movement under conditions of turbulence we could expect the curve to be rough and nonlinear (which means it is not possible to derive the differential equations for describing this curve, thus making predictions concerning the fluid's movement very difficult if not impossible). However, viscous fluids, for which the Navier-Stokes equations are formulated, are able to smooth out or overcome the ripples, eddies and whorls of turbulence, as mentioned above, the more viscous the fluids the more able they are in doing so. Hence, the more viscous the fluid the more successful the Navier-Stokes equations should be in describing the motion of the fluid.

If turbulence or chaos could be predicted by a mathematical equation or equations then it is not really turbulence or chaos. One should bear in mind that chaos, as the term implies, results in disorder, having no discernable pattern, confusion and puzzlement, which is the contrary of the state of being orderly, having an obvious pattern, being deterministic and being predictable. Nevertheless, whatever method we adopt for describing turbulence or chaos we still need to confirm its validity through experiments, just as the validity of the Navier-Stokes equations for two-dimensional fluid motions has been confirmed by physical experiments. Thus, we should first proceed with the physical experiments to get a better understanding of turbulence or chaos in incompressible fluids.

Fluid Viscosity And Turbulence
The more viscous (sticky or gooey) the incompressible fluid is the less able it is to flow or move freely, the less runny it is, and, as stated above, the more able it is to overcome or smooth out the whorls, eddies

and ripples of turbulence. A fluid might be so viscous and its flow so restricted that it appears almost like a solid. Such fluids which come to mind are, e.g., paints and polymer solutions. What does all this imply?

It is evident that there is a threshold or cut-off point in the viscosity of a fluid at and above which turbulence does not affect it much, which means that the effects of turbulence are only marked if a fluid has a viscosity that is less than this threshold or cut-off point. At or beyond this threshold, this cut-off point, one could expect the Navier-Stokes equations not to fare too badly.

How would the introduction of a large initial force affect a viscous fluid whose viscosity is at or above this threshold? If a large initial force is introduced, this fluid could be expected to remain "lumpy" instead of moving more freely or becoming more runny, i.e., turbulence could not be expected to appear.

The opposite is true below this threshold. Below this cut-off point or threshold, i.e., if the fluid is not too viscous, with the initial force large enough, the fluid's Reynolds number stands a chance of exceeding 2000 to 3000, with turbulence setting in, and whorls, eddies and ripples forming in the fluid. The Navier-Stokes equations would encounter difficulties.

Two-Dimensional Flows And Their Solutions
Solutions of the Navier-Stokes equations result in velocity vectors, q, and pressures, p, which satisfy both the momentum equations and the continuity equation. If one were given such a combination, [q, p], one could check whether it constitutes a solution by substitution into the equations. How such a solution is found is something else. Any general step leading to this goal is helpful. For two-dimensional flows, it is possible to get rid of the continuity equation from the system of equations by using only functions which satisfy the continuity equation. This elimination is a formal step towards a solution. The functions which affect this elimination are the stream functions.

A flow could be defined as two-dimensional when its description in Cartesian coordinates shows no z-component of the velocity and no dependence on the z-coordinate. A flow like this could be described in the z = 0 plane, with the velocity vector and the streamlines lying in this plane. Moreover, the z = C planes, which are parallel to the z = 0 plane, display a flow pattern which is identical to that in the z = 0 plane. The z = 0 plane is called the representative plane.

Three-Dimensional Flows And Practical Solutions
Historically, the dynamics of turbulence or chaos is the corollary of Poincare's "three-body" problem concerning planetary motions which Poincare found to be very complex and unsolvable.

Geometrically, turbulence in fluids cannot be described by the use of the "Poincare section", i.e., there is no periodic solution for it. (In the use of the "Poincare section", for there to be a periodic solution, a circular curve must return to the section at its exact starting point. In the condition of turbulence the curve would not return to its exact starting point and there is therefore no periodic solution.)

Our solutions here are to be implemented as much as possible in the spirit of the Navier-Stokes equations that pertain to the "behaviour" of an incompressible viscous fluid in three dimensions (x, y, z). We aim to overcome the obstacle encountered by the Navier-Stokes equations in the three-dimensional case.

At least two methods of experimentation could be carried out to get a good picture of turbulence in incompressible fluids. This "picture" is however only an "approximation", with a "factor of accuracy" - which comprises an upper limit and a lower limit. The first method involves bouncing laser beams off a reflective strip that is immersed in the turbulent viscous fluid, making use of the Doppler effect, to determine the velocity of the strip, and, thus, the velocity of the fluid. However, this method only provides data obtained from one point, axis or dimension, though we could extrapolate the data for the other two dimensions to get a three-dimensional picture. The second method involves the simulation of fluid motions in conditions of turbulence with powerful computers. The fact that turbulence or chaos is unpredictable implies that the result of fluid motions in turbulent conditions is anybody's guess. The guesses could vary widely in scope but each of them has a "probability" of being correct. A statistical method of analysing fluid motions in conditions of turbulence would be strongly recommended, which is in some way similar in principle to the above-mentioned statistical mechanics. Though statistical mechanics could be used as an example for describing the behaviours of incompressible fluids under conditions of turbulence (e.g., the way it is being used to predict the behaviours and positions of quantum particles - only in a probabilistic manner), applying its methods is bound to face practical difficulties. (How do we assign "probability ratings" to the behaviours of quantum particles? What are the bases for such "ratings"? Are the "ratings" based on experimental proofs or data, which should be the case?) A suggestion is to adopt a method of analysing

fluid flows under conditions of turbulence based on statistical data obtained through actual, rigorous experiments, the utilisation of statistical and interpolation/extrapolation methods, and, sound common sense or logic.

The second method, which the author strongly recommends, involves carrying out an experiment through computer simulation. Computer simulation is very powerful and is commonly used today. Instead of carrying out an experiment involving real fluids and real turbulence, which would be cumbersome, we would simulate this experiment with powerful computers, which would be more practicable and would cut down costs and save time considerably. (In computer simulations electrons are actually being utilised to represent the object or objects being simulated and these simulations could be carried out in three dimensions - in this case viscous, incompressible fluid in turbulence with Reynolds numbers above 3000 being simulated in three dimensions as is described below.) As the flow of the fluid is completely turbulent when the Reynolds number exceeds about 2000 to 3000, we could via our powerful computers simulate fluid flows under conditions of turbulence for Reynolds numbers of, e.g., 3050, 3300, 3550, 3800, 4050, 4300, 4550, 4800 and 5050 respectively and thereby obtain the velocity results of fluid motion (by the process of iteration) for each of these Reynolds numbers.

We would use three powerful computers or work stations which work at very high speeds for these computer simulations. For each of the above-mentioned nine Reynolds numbers (3050, 3300, 3550, 3800, 4050, 4300, 4550, 4800 and 5050) the three computers or work stations would carry out the following. The first computer would perform the simulation of fluid movement under conditions of turbulence and the fluid's velocity at a point, P, in the fluid would be measured in the x direction or axis (front view or plane) by a probe at this point, P, and recorded. The second computer would perform the same simulation but at a different plane or view, say, the side view (y), and the fluid's velocity at point P in the fluid would be measured in the y direction or axis (side view or plane) by a probe at point P and recorded. The third computer would perform the same simulation at a yet different plane or view, now, the top view (z), and the fluid's velocity at point P in the fluid would be measured in the z direction or axis (top view or plane) by a probe at point P and recorded. Hence, we now have the fluid's velocities (v = m/s) at each of the three planes or axes, x, y, z, for the same Reynolds number, which are as follows:-

(i)   The fluid's velocity in the x axis - v (x)
(ii)   The fluid's velocity in the y axis - v (y)
(iii) The fluid's velocity in the z axis - v (z)

For each of the nine Reynolds numbers, the three computers or work stations would each carry out (iterate) the simulations, say, 1 million times (the more simulations the better) to produce respectively 1 million values for v (x), 1 million values for v (y) and 1 million values for v (z).

(The principles involved in measuring the velocities of the fluid in the x, y, z directions or axes are explained here. The probes mentioned above should be made of thin material and plate-like and should offer minimal resistance against the flow of the fluid, i.e., offer minimal interference with the flow of the fluid. For each axis or plane, the flat surface of the probe should face the direction of the axis. The velocity of the fluid at each axis, plane or dimension is a function of the fluid pressure on the flat surface of the probe, the higher the fluid pressure the higher the velocity of the fluid and vice versa. The probe in each axis or plane should be able to measure both positive (+) velocities and negative (-) velocities, i.e., velocities in the opposite directions.)

This process for all the above-mentioned nine Reynolds numbers would be carried out, giving a total of 27 million simulations, and, 9 million values for v (x), 9 million values for v (y) and 9 million values for v (z).

We next apply the statistical method of time-series analysis (which is a method of statistical analysis designed to eliminate seasonal variations in trends, used for forecasting and prediction, but implemented with some modification here). For each of the above-mentioned nine Reynolds numbers, we compute (with the aid of the computer) the moving averages (m) for the 1 million v (x)'s, 1 million v (y)'s and 1 million v (z)'s, regardless of whether v (x), v (y) and v (z) are positive (+) or negative (-), by only adding (no subtracting) and dividing, as follows:-

(i) $v_1 (x) + v_2 (x) \div 2 = m_1 (x)$

(ii) $v_1 (x) + v_2 (x) + v_3 (x) \div 3 = m_2 (x)$

(iii) $v_1 (x) + v_2 (x) + v_3 (x) + v_4 (x) \div 4 = m_3 (x)$

.

.

.

.

(*q) $v_1 (x) + v_2 (x) + v_3 (x) + v_4 (x) + \ldots\ldots + v_{q + 1} (x) \div q + 1 = m_q (x)$

(*: q = 999,999)

whereby $m_1 (x)$, $m_2 (x)$, $m_3 (x) \ldots m_q (x)$ are the 999,999 moving averages (m) for v (x).

(The same applies to the 1 million v (y)'s and the 1 million v (z)'s.)

Thus, we have 999,999 (q) moving averages (m) each for each of the 1 million v (x)'s, 1 million v (y)'s and 1 million v (z)'s for each of the above-mentioned nine Reynolds numbers (giving a total of 26,999,973 moving averages (m) for all the above-mentioned nine Reynolds numbers).

For each of the above-mentioned nine Reynolds numbers, substituting 999,999 with q, we get the qth. moving averages (m_q) for
v (x), v (y) and v (z), which are actually each respectively the 999,999th. average velocity of the fluid in each of the three dimensions or axes, x, y, z, which are as follows:-

(i) $m_q (x)$

(ii) $m_q (y)$

(iii) $m_q (z)$

Upon the completion of all the computer simulations and computations of the moving averages (m), we would produce a statistical table with the fluid velocities, m_q (x), m_q (y) and m_q (z), in the three dimensions or axes, x (front view), y (side view) and z (top view), for each of the nine Reynolds numbers: 3050, 3300, 3550, 3800, 4050, 4300, 4550, 4800 and 5050. For each of the three velocities, m_q (x), m_q (y) and m_q (z), for each of these nine Reynolds numbers, we would include an upper limit of accuracy, u (x), u (y) and u (z) (each of which is respectively the largest moving average (m) that has been obtained through carrying out the 1 million simulations for each of the three dimensions or axes, x, y, z), and, a lower limit of accuracy, l (x), l (y) and l (z) (each of which is respectively the smallest moving average (m) which has been obtained through carrying out the 1 million simulations for each of the three dimensions or axes, x, y, z). For example, for each of these nine Reynolds numbers we would have the following velocities with their respective upper and lower limits of accuracy:-

(i)   x dimension or axis - m_q (x), u (x), l (x)
(ii)   y dimension or axis - m_q (y), u (y), l (y)
(iii)   z dimension or axis - m_q (z), u (z), l (z)

However, for the Reynolds numbers between these nine Reynolds numbers, i.e., the "intermediate" Reynolds numbers (such as 3200, 4400 and 4950, e.g.), we would obtain the velocity results with their respective upper and lower limits of accuracy, i.e., the m_q's and their respective u's and l's, through interpolation (i.e., estimate the velocity results and their respective upper and lower limits of accuracy - some methods of interpolation are the Lagrange interpolation and the Gregory-Newton interpolation).

With the above-mentioned statistical data for the nine Reynolds numbers we would apply the rules of vector calculus to obtain for each of these nine Reynolds numbers the "resultant" velocity of the fluid in three dimensions or axes (x, y, z) and its upper and lower limits of accuracy. For example, we could obtain the "resultant" velocity and its upper and lower limits of accuracy for unidirectional fluid flow in the three dimensions or axes (x, y, z) as follows:-

m_q (x, y, z), u (x, y, z), l (x, y, z) = + m_q (x) + m_q (y) + m_q (z); + u (x) + u (y) + u (z); + l (x) + l (y) + l (z) treated as negative (-) if more than 50% of the 1 million v (z)'s are negative (-).

Hence, when more than 50% of the 1 million v (x)'s, v (y)'s or v (z)'s are negative (-), i.e., move in the opposite direction, and their moving average (m_q) is hence treated as negative (-), we have to subtract m_q (x), m_q (y) or m_q (z), e.g., if m_q (z) is negative (-), we have to compute the "resultant" velocity m_q (x, y, -z) as follows:-

m_q (x, y, - z) = + m_q (x) + m_q (y) - m_q (z), i.e., subtract m_q (z) from + m_q (x) + m_q (y), with its upper and lower limits of accuracy computed as follows:

(i) $u(x, y, -z) = + u(x) + u(y) - u(z)$, where $u(x)$ is the largest moving average ($m(x)$) among the 999,999 (q) moving averages (m) for the 1 million $v(x)$'s, $u(y)$ is the largest moving average ($m(y)$) among the 999,999 (q) moving averages (m) for the 1 million $v(y)$'s, and, $u(z)$ is the largest moving average ($m(z)$) among the 999,999 (q) moving averages (m) for the 1 million $v(z)$'s.

(ii) $l(x, y, -z) = + l(x) + l(y) - l(z)$, where $l(x)$ is the smallest moving average ($m(x)$) among the 999,999 (q) moving averages (m) for the 1 million $v(x)$'s, $l(y)$ is the smallest moving average ($m(y)$) among the 999,999 (q) moving averages (m) for the 1 million $v(y)$'s, and, $l(z)$ is the smallest moving average ($m(z)$) among the 999,999 (q) moving averages (m) for the 1 million $v(z)$'s.

"Velocity" diagrams (or vector diagrams) could be produced for all the *possible* "resultant" velocities indicated below, showing their directions of flow, which vary:-

(i) For $+ m_q(x) + m_q(y) + m_q(z)$; $+ u(x) + u(y) + u(z)$; $+ l(x) + l(y) + l(z)$, we get the "resultant" velocity $m_q(x, y, z)$, $u(x, y, z)$, $l(x, y, z)$.

(ii) For $- m_q(x) - m_q(y) - m_q(z)$; $- u(x) - u(y) - u(z)$; $- l(x) - l(y) - l(z)$, we get the "resultant" velocity $m_q(-x, -y, -z)$, $u(-x, -y, -z)$, $l(-x, -y, -z)$.

(iii) For $- m_q(x) - m_q(y) + m_q(z)$; $- u(x) - u(y) + u(z)$; $- l(x) - l(y) + l(z)$, we get the "resultant" velocity $m_q(-x, -y, z)$, $u(-x, -y, z)$, $l(-x, -y, z)$ (where $- m_q(x) - m_q(y) > + m_q(z)$).

(iv) For $+ m_q(x) - m_q(y) - m_q(z)$; $+ u(x) - u(y) - u(z)$; $+ l(x) - l(y) - l(z)$, we get the "resultant" velocity $m_q(x, -y, -z)$, $u(x, -y, -z)$, $l(x, -y, -z)$ (where $- m_q(y) - m_q(z) > + m_q(x)$).

(v) For $- m_q(x) + m_q(y) - m_q(z)$; $- u(x) + u(y) - u(z)$; $- l(x) + l(y) - l(z)$, we get the "resultant" velocity $m_q(-x, y, -z)$, $u(-x, y, -z)$, $l(-x, y, -z)$ (where $- m_q(x) - m_q(z) > + m_q(y)$).

(vi) For $- m_q(x) + m_q(y) + m_q(z)$; $- u(x) + u(y) + u(z)$; $- l(x) + l(y) + l(z)$, we get the "resultant" velocity $m_q(-x, y, z)$, $u(-x, y, z)$, $l(-x, y, z)$ (where $+ m_q(y) + m_q(z) > - m_q(x)$).

(vii) For $+ m_q(x) - m_q(y) + m_q(z)$; $+ u(x) - u(y) + u(z)$; $+ l(x) - l(y) + l(z)$, we get the "resultant" velocity $m_q(x, -y, z)$, $u(x, -y, z)$, $l(x, -y, z)$ (where $+ m_q(x) + m_q(z) > - m_q(y)$).

(viii) For $+ m_q(x) + m_q(y) - m_q(z)$; $+ u(x) + u(y) - u(z)$; $+ 1(x) + 1(y) - 1(z)$, we get the "resultant" velocity $m_q(x, y, - z)$, $u(x, y, - z)$, $1(x, y, - z)$ (where $+ m_q(x) + m_q(y) > - m_q(z)$).

So far, in the above "velocity" scenarios, for those with either, one negative velocity $(- m_q)$ and two positive velocities $(+ m_q)$, or, one positive velocity $(+ m_q)$ and two negative velocities $(- m_q)$, it is assumed that for the former the two positive velocities $(+ m_q)$ combined together are larger than the one negative velocity $(- m_q)$, and, for the latter it is assumed that the two negative velocities $(- m_q)$ combined together are larger than the one positive velocity $(+ m_q)$. Now, what happens when for the former, the two positive velocities $(+ m_q)$ combined together are equal to the one negative velocity $(- m_q)$, and, for the latter, the two negative velocities $(- m_q)$ combined together are equal to the one positive velocity $(+ m_q)$? In each of these cases, the "resultant" velocity $(m_q(x, y, z), u(x, y, z), 1(x, y, z))$ would be equal to zero, i.e., there would be no "resultant" velocity.

What happens when the negative velocity, $- m_q(x)$, $- m_q(y)$, or, $- m_q(z)$, is larger than the other two positive velocities combined together, e.g., $- m_q(x) > + m_q(y) + m_q(z)$, $- m_q(y) > + m_q(x) + m_q(z)$, or, $- m_q(z) > + m_q(x) + m_q(y)$? For such cases we have the following "resultant" velocities, whose directions of flow differ from each other:-

(i) For $- m_q(x) + m_q(y) + m_q(z)$; $- u(x) + u(y) + u(z)$; $- 1(x) + 1(y) + 1(z)$, where $- m_q(x) > + m_q(y) + m_q(z)$, we get the "resultant" velocity $m_q(- x, y, z)$, $u(- x, y, z)$, $1(- x, y, z)$.

(ii) For $+ m_q(x) - m_q(y) + m_q(z)$; $+ u(x) - u(y) + u(z)$; $+ 1(x) - 1(y) + 1(z)$, where $- m_q(y) > + m_q(x) + m_q(z)$, we get the "resultant" velocity $m_q(x, - y, z)$, $u(x, - y, z)$, $1(x, - y, z)$.

(iii) For $+ m_q(x) + m_q(y) - m_q(z)$; $+ u(x) + u(y) - u(z)$; $+ 1(x) + 1(y) - 1(z)$, where $- m_q(z) > + m_q(x) + m_q(y)$, we get the "resultant" velocity $m_q(x, y, - z)$, $u(x, y, - z)$, $1(x, y, - z)$.

What happens then when the positive velocity, $+ m_q(x)$, $+ m_q(y)$, or, $+ m_q(z)$ is larger than the other two negative velocities combined together, e.g., $+ m_q(x) > - m_q(y) - m_q(z)$, $+ m_q(y) > - m_q(x) - m_q(z)$, or, $+ m_q(z) > - m_q(x) - m_q(y)$? For such cases we have the following "resultant" velocities, whose directions of flow differ from each other:-

(i) For $+ m_q(x) - m_q(y) - m_q(z)$; $+ u(x) - u(y) - u(z)$; $+ 1(x) - 1(y) - 1(z)$, where $+ m_q(x) > - m_q(y) - m_q(z)$, we get the "resultant" velocity $m_q(x, - y, - z)$, $u(x, - y, - z)$, $1(x, - y, - z)$.

(ii) For $- m_q(x) + m_q(y) - m_q(z)$; $- u(x) + u(y) - u(z)$; $- 1(x) + 1(y) - 1(z)$, where $+ m_q(y) > - m_q(x) - m_q(z)$, we get the "resultant" velocity $m_q(- x, y, - z)$, $u(- x, y, - z)$, $1(- x, y, - z)$.

(iii) For $- m_q (x) - m_q (y) + m_q (z)$; $- u (x) - u (y) + u (z)$; $- l (x) - l (y) + l (z)$, where $+ m_q (z) > - m_q (x) - m_q (y)$, we get the "resultant" velocity $m_q (- x, - y, z)$, $u (- x, - y, z)$, $l (- x, - y, z)$.

For the three velocities in the x, y, z dimensions or axes, i.e., $m_q (x)$, $m_q (y)$ and $m_q (z)$, there are 14 possible "resultant" velocities, excluding the null "resultant" velocities, of which there are six possible cases (which are not expected to be likely to occur in turbulence), as shown above. Each of these "resultant" velocities would have an upper limit of accuracy ($u (x, y, z)$) and a lower limit of accuracy ($l (x, y, z)$). For fluid velocities ($m_q (x)$, $m_q (y)$, or, $m_q (z)$) in the opposite direction their values are negative. We could expect turbulence to be characterized by any of these 14 "resultant" velocities.

Nevertheless, there is a possibility that one or more of the velocities, $m_q (x)$, $m_q (y)$ and $m_q (z)$, might be equal to zero, though the chances of this occurring in turbulence might be remote:-

(i) If, e.g., $m_q (z) = 0$, then $m_q (x, y, z = 0)$, $u (x, y, z = 0)$, $l (x, y, z = 0)$ would be a "resultant" velocity in two dimensions or axes only, with the fluid moving in the two dimensions or axes, x and y, only.

(ii) For $m_q (x) = 0$, we get the "resultant" velocity $m_q (x = 0, y, z)$, $u (x = 0, y, z)$, $l (x = 0, y, z)$.
(iii) For $m_q (y) = 0$, we get the "resultant" velocity $m_q (x, y = 0, z)$, $u (x, y = 0, z)$, $l (x, y = 0, z)$.

As in the three-dimensional case above, there is the possibility in the two-dimensional case that one or both of the two velocities are negative or positive velocities. If, in the case whereby one of the two velocities is negative while the other is positive, the negative velocity is equal to the positive velocity (of which there are six possible cases), then the "resultant" velocity is null (zero). The following are the possible "resultant" velocities:-

(i) For $m_q (z) = 0$, we get the "resultant" velocity $m_q (x, - y, z = 0)$, $u (x, - y, z = 0)$, $l (x, - y, z = 0)$, when $m_q (x)$ is positive while $m_q (y)$ is negative, whereby $- m_q (y) > + m_q (x)$.

(ii) For $m_q (z) = 0$, we get the "resultant" velocity $m_q (x, - y, z = 0)$, $u (x, - y, z = 0)$, $l (x, - y, z = 0)$, when $m_q (x)$ is positive while $m_q (y)$ is negative, whereby $+ m_q (x) > - m_q (y)$.

(iii) For $m_q (z) = 0$, we get the "resultant" velocity $m_q (- x, y, z = 0)$, $u (- x, y, z = 0)$, $l (- x, y, z = 0)$, when $m_q (y)$ is positive while $m_q (x)$ is negative, whereby $- m_q (x) > + m_q (y)$.

(iv) For $m_q (z) = 0$, we get the "resultant" velocity $m_q (- x, y, z = 0)$, $u (- x, y, z = 0)$, $l (- x, y, z = 0)$, when $m_q (y)$ is positive while $m_q (x)$ is negative, whereby $+ m_q (y) > - m_q (x)$.

(v) For $m_q(z) = 0$, we get the "resultant" velocity $m_q(-x, -y, z = 0)$, $u(-x, -y, z = 0)$, $l(-x, -y, z = 0)$, when $m_q(x)$ and $m_q(y)$ are both negative.

(vi) For $m_q(z) = 0$, we get the "resultant" velocity $m_q(x, y, z = 0)$, $u(x, y, z = 0)$, $l(x, y, z = 0)$, when $m_q(x)$ and $m_q(y)$ are both positive.

(vii) For $m_q(x) = 0$, we get the "resultant" velocity $m_q(x = 0, -y, z)$, $u(x = 0, -y, z)$, $l(x = 0, -y, z)$, when $m_q(z)$ is positive while $m_q(y)$ is negative, whereby $-m_q(y) > +m_q(z)$.

(viii) For $m_q(x) = 0$, we get the "resultant" velocity $m_q(x = 0, -y, z)$, $u(x = 0, -y, z)$, $l(x = 0, -y, z)$, when $m_q(z)$ is positive while $m_q(y)$ is negative, whereby $+m_q(z) > -m_q(y)$.

(ix) For $m_q(x) = 0$, we get the "resultant" velocity $m_q(x = 0, y, -z)$, $u(x = 0, y, -z)$, $l(x = 0, y, -z)$, when $m_q(y)$ is positive while $m_q(z)$ is negative, whereby $-m_q(z) > +m_q(y)$.

(x) For $m_q(x) = 0$, we get the "resultant" velocity $m_q(x = 0, y, -z)$, $u(x = 0, y, -z)$, $l(x = 0, y, -z)$, when $m_q(y)$ is positive while $m_q(z)$ is negative, whereby $+m_q(y) > -m_q(z)$.

(xi) For $m_q(x) = 0$, we get the "resultant" velocity $m_q(x = 0, -y, -z)$, $u(x = 0, -y, -z)$, $l(x = 0, -y, -z)$, when $m_q(y)$ and $m_q(z)$ are both negative.

(xii) For $m_q(x) = 0$, we get the "resultant" velocity $m_q(x = 0, y, z)$, $u(x = 0, y, z)$, $l(x = 0, y, z)$, when $m_q(y)$ and $m_q(z)$ are both positive.

(xiii) For $m_q(y) = 0$, we get the "resultant" velocity $m_q(-x, y = 0, z)$, $u(-x, y = 0, z)$, $l(-x, y = 0, z)$, when $m_q(z)$ is positive while $m_q(x)$ is negative, whereby $-m_q(x) > +m_q(z)$.

(xiv) For $m_q(y) = 0$, we get the "resultant" velocity $m_q(-x, y = 0, z)$, $u(-x, y = 0, z)$, $l(-x, y = 0, z)$, when $m_q(z)$ is positive while $m_q(x)$ is negative, whereby $+m_q(z) > -m_q(x)$.

(xv) For $m_q(y) = 0$, we get the "resultant" velocity $m_q(x, y = 0, -z)$, $u(x, y = 0, -z)$, $l(x, y = 0, -z)$, when $m_q(x)$ is positive while $m_q(z)$ is negative, whereby $-m_q(z) > +m_q(x)$.

(xvi) For $m_q(y) = 0$, we get the "resultant" velocity $m_q(x, y = 0, -z)$, $u(x, y = 0, -z)$, $l(x, y = 0, -z)$, when $m_q(x)$ is positive while $m_q(z)$ is negative, whereby $+m_q(x) > -m_q(z)$.

(xvii) For $m_q(y) = 0$, we get the "resultant" velocity $m_q(-x, y = 0, -z)$, $u(-x, y = 0, -z)$, $l(-x, y = 0, -z)$, when $m_q(x)$ and $m_q(z)$ are both negative.

(xviii) For m_q (y) = 0, we get the "resultant" velocity m_q (x, y = 0, z), u (x, y = 0, z), l (x, y = 0, z), when m_q (x) and m_q (z) are both positive.

(xix) For m_q (x) = 0 and m_q (y) = 0, we get the "resultant" velocity m_q (x = 0, y = 0, z), u (x = 0, y = 0, z), l (x = 0, y = 0, z) in one dimension or axis, which is positive.

(xx) For m_q (x) = 0 and m_q (y) = 0, we get the "resultant" velocity m_q (x = 0, y = 0, - z), u (x = 0, y = 0, - z), l (x = 0, y = 0, - z) in one dimension or axis, which is negative.

(xxi) For m_q (x) = 0 and m_q (z) = 0, we get the "resultant" velocity m_q (x = 0, y, z = 0), u (x = 0, y, z = 0), l (x = 0, y, z = 0) in one dimension or axis, which is positive.

(xxii) For m_q (x) = 0 and m_q (z) = 0, we get the "resultant" velocity m_q (x = 0, - y, z = 0), u (x = 0, - y, z = 0), l (x = 0, - y, z = 0) in one dimension or axis, which is negative.

(xxiii) For m_q (y) = 0 and m_q (z) = 0, we get the "resultant" velocity m_q (x, y = 0, z = 0), u (x, y = 0, z = 0), l (x, y = 0, z = 0) in one dimension or axis, which is positive.

(xxiv) For m_q (y) = 0 and m_q (z) = 0, we get the "resultant" velocity m_q (- x, y = 0, z = 0), u (- x, y = 0, z = 0), l (- x, y = 0, z = 0) in one dimension or axis, which is negative.

After all the simulations and computations of moving averages (m) for the 27 million fluid velocities (v) in the x, y, z dimensions or axes, we firstly obtain a table, which would act as a rough statistical guide for fluid flow under conditions of turbulence, with the following statistical data, for each of the nine Reynolds numbers (3050, 3300, 3550, 3800, 4050, 4300, 4550, 4800 and 5050):-

(1) (i) The fluid's average velocity in the x dimension or axis - m_q (x)
     (ii) This fluid average velocity's upper limit of accuracy - u (x)
         (i.e., largest moving average (m) for v (x))
     (iii) This fluid average velocity's lower limit of accuracy - l (x)

         (i.e., smallest moving average (m) for v (x))

(2) (i) The fluid's average velocity in the y dimension or axis - m_q (y)
     (ii) This fluid average velocity's upper limit of accuracy - u (y)
         (i.e., largest moving average (m) for v (y))
     (iii) This fluid average velocity's lower limit of accuracy - l (y)

(i.e., smallest moving average (m) for v (y))

(3) (i) The fluid's average velocity in the z dimension or axis - m_q (z)
   (ii) This fluid average velocity's upper limit of accuracy - u (z)
     (i.e., largest moving average (m) for v (z))
  (iii) This fluid average velocity's lower limit of accuracy - l (z)
     (i.e., smallest moving average (m) for v (z))

(4) (i) The fluid's "resultant" velocity in the x, y, z dimensions or axes - m_q (x, y, z)
     (i.e., +/- m_q (x) +/- m_q (y) +/- m_q (z))
  (ii) This fluid "resultant" velocity's upper limit of accuracy - u (x, y, z)
     (i.e., +/- m (x) +/- m (y) +/- m (z))
  (iii) This fluid "resultant" velocity's lower limit of accuracy - l (x, y, z)
     (i.e., +/- m (x) +/- m (y) +/- m (z))

The "velocity" diagrams depicting the various "resultant" velocities and the directions of flow should also be included, a total of 27 diagrams, three "velocity" diagrams for each of the nine Reynolds numbers - one for m_q (x, y, z), one for u (x, y, z) and one for l (x, y, z). This table may be titled "Statistical Data Of An Incompressible Fluid's Velocities For The Nine Reynolds Numbers: 3050, 3300, 3550, 3800, 4050, 4300, 4550, 4800 And 5050".

For the "intermediate" Reynolds numbers, e.g., 3200, 4400 and 4950, we would interpolate the fluid's respective velocities and their upper and lower limits of accuracy with the above-mentioned statistical table (Statistical Data Of An Incompressible Fluid's Velocities For The Nine Reynolds Numbers: 3050, 3300, 3550, 3800, 4050, 4300, 4550, 4800 And 5050) and the complementary table (Velocity Results Of The 27 Million Simulations In The X, Y, Z Dimensions) described below as guides, i.e., estimate them - some methods of interpolation that could be adopted are the Lagrange interpolation and the Gregory-Newton interpolation. For velocities for Reynolds numbers outside the range of the above nine Reynolds numbers (3050 to 5050), e.g., for Reynolds numbers 2500 and 5200, we would extrapolate them, including their upper and lower limits of accuracy, using the same tables as guides.

With the above techniques (and with proper simulations) we thus have a statistical table of fluid flow velocity results (Statistical Data Of An Incompressible Fluid's Velocities For The Nine Reynolds Numbers: 3050, 3300, 3550, 3800, 4050, 4300, 4550, 4800 And 5050) under conditions of turbulence (with their respective ranges of possibilities or probabilities, i.e., their respective upper limits and lower limits of accuracy) for the various Reynolds numbers which are more or less rigorously based on actual experiments (simulations in this case), a table which is somewhat like, e.g., the statistical tables for t-distribution and chi-squared distribution. In the above-mentioned experiments (or simulations), we could expect more accurate results with more simulations being carried out (more data being collected), the more the simulations carried out the more accurate the results are likely to be. This represents a practical, logical way of roughly approximating the velocities of fluid motions under conditions of turbulence (in three dimensions) at the various Reynolds numbers. Given a fluid velocity (m_q (x, y, z), u (x, y, z), l (x, y, z)) from the above-mentioned statistical table, we could in principle tell (or predict) for a particular Reynolds number the position (position after distance travelled, d_2, relative to the position at $d_1 = 0$) of an object, e.g., an aluminium strip, which could be taken to represent a section of the fluid, carried along by the fluid in motion, at a point of time, $t_2 > 0$, with the initial time being $t_1 = 0$, or, the distance ($\{d_2 > 0\}$ - $\{d_1 = 0\}$) this object travelled at a point of time, $t_2 > 0$, with the initial time being $t_1 = 0$, by computation with the equation: d_2 - $d_1 = (m_q$ (x, y, z), u (x, y, z), l (x, y, z)) x (t_2 - t_1). However, in this instance, wherein turbulence rules, this prediction of position or distance is not expected to be accurate (but should be regarded only as a rough approximation) and should be subject to the statistical rule of "probability" - in this case the upper limit ($\{u$ (x, y, z)$\}$ x $\{t_2$ - $t_1\}$) and the lower limit ($\{l$ (x, y, z)$\}$ x $\{t_2$ - $t_1\}$) of the accuracy of the result obtained, as indicated by the above-mentioned statistical table. The "velocity" diagrams in the statistical table would act as a guide with regards to the positioning of the object and the direction it would travel. Here the movement of the object, which is the aluminium strip, in effect represents the movement of the fluid which carries it along. However, the data reflected in this first statistical table (Statistical Data Of An Incompressible Fluid's Velocities For The Nine Reynolds Numbers: 3050, 3300, 3550, 3800, 4050, 4300, 4550, 4800 And 5050) should be regarded only as a rough approximation. With the further assistance of the complementary table (Velocity Results Of The 27 Million Simulations In The X, Y, Z Dimensions) described below this approximation could be refined.

With the data from the above-mentioned statistical table (Statistical Data Of An Incompressible Fluid's Velocities For The Nine Reynolds Numbers: 3050, 3300, 3550, 3800, 4050, 4300, 4550, 4800 And 5050) it is possible to plot a curve for the fluid flow velocity results (m_q (x, y, z)) for the above-mentioned nine Reynolds numbers. If this curve is smooth and linear, which is unlikely, from the gradient of the slope of this curve the differential equations for this curve could be derived, whereby forecasts or predictions would be possible. However, if this curve is rough and nonlinear, which is likely to be the case, then these differential equations would not be obtainable, and, we would have to rely on the above-mentioned table of fluid flow velocity results (Statistical Data Of An Incompressible Fluid's Velocities For The Nine Reynolds Numbers: 3050, 3300, 3550, 3800, 4050, 4300, 4550, 4800 And 5050), and, the complementary table (Velocity Results Of The 27 Million Simulations In The X, Y, Z Dimensions) described below, as statistical guides for our approximation (including interpolation and extrapolation) of velocity results for fluid flows at the various Reynolds numbers. Both these statistical tables with their "velocity" diagrams should complement one another and should be a great help for this approximation process, the more copious the data available there the more effective the approximation should be. It is thus possible to approximate not only fluid velocities but directions of fluid motions as well (with the aid of the "velocity" diagrams). We could continue to plot graphs or curves with these statistical data and attempt to interpolate or extrapolate with them, looking out for trends or patterns, and arrive at some forecasts or predictions, which is expected to be a difficult task. By doing this we could at least get a "feel" or intuitive understanding of the whole situation, which should make the job of forecasting or predicting the outcome easier, an evidently challenging undertaking.

The above-mentioned statistical table, Statistical Data Of An Incompressible Fluid's Velocities For The Nine Reynolds Numbers: 3050, 3300, 3550, 3800, 4050, 4300, 4550, 4800 And 5050, presents the average velocities for the nine Reynolds numbers. These average velocities each represents the velocity "trend" for each of the nine Reynolds numbers. However, carrying out the approximation for the respective velocities for the respective Reynolds numbers, wherein complete turbulence is involved, might not be that easy. It should be remembered that the Navier-Stokes equations fare very badly when turbulence sets in, when a viscous, incompressible fluid behaves in an unpredictable, irregular, chaotic, and, "nonlinear" manner. To complement the first statistical table (Statistical Data Of An Incompressible Fluid's Velocities For The Nine Reynolds Numbers: 3050, 3300, 3550, 3800, 4050, 4300, 4550, 4800 And 5050), it would be appropriate to have a proper tabulation of all the 27 million fluid velocities (v) in the x, y, z dimensions or axes obtained through the simulations in another table (or booklet), which may be titled "Velocity Results Of The 27 Million Simulations In The X, Y, Z Dimensions," which shows whole ranges of velocities (minimum velocity, maximum velocity, and, all the velocities between the minimum and the maximum, a total of 27 million velocities (v's), three million velocities (v's) for each of the nine Reynolds numbers), with their order of listing in this table (or booklet) exactly the same as the order they appeared in during the simulations, with indications whether they are positive (+) or negative (-) velocities, and, the 999,999 (q) moving averages (m) each for each of the 1 million v (x)'s, 1 million v (y)'s and 1 million v (z)'s for each of the nine Reynolds numbers (giving a total of 26,999,973 moving averages (m) for all the nine Reynolds numbers) included. The "velocity" diagrams described above, which would also be helpful as guides, should be incorporated, e.g., as an appendix, in this table (or booklet). With this additional table and its "velocity" diagrams acting as a further guide, the task of approximating (including interpolating and extrapolating) the velocities for the various Reynolds numbers (for turbulence) would be made easier. With these two tables there are now more or less solid and realistic data to carry out the approximation (including interpolation and extrapolation) - for both fluid velocities, and, directions of fluid motions. The approximations could also include probabilities of occurrence (an example of which is presented below). However, sound common sense, good intuition and a sharp eye for patterns and details are important for the approximation (including interpolation and extrapolation). Since turbulence is a much complex phenomenon, the approximation (including interpolation and extrapolation) should be carried out with great care and patience. There are two choices here now. We could supplant the approximation method of the Navier-Stokes equations with the above-mentioned statistical method. Alternatively, we could use this statistical method as a complement to the Navier-Stokes equations, whereby we might have the "best of both worlds".

Conclusion

There is a possibility, however remote, that turbulence or chaos when viewed en masse, on a very large scale, would appear to be smooth, present some sort of pattern or appear to have some order. According to the precepts of fractal geometry, a relatively new branch of mathematics pioneered by Benoit Mandelbrot, phenomena which appear random, when viewed en masse, display some orderliness and pattern, which could be termed "fractal". Thus, a new mathematical technique for describing the flow of an incompressible viscous fluid in three dimensions, i.e., in turbulence, a probably statistical one involving large samples of data, like the method described just above, is a logical step.

As for the case of computer simulation, it is becoming more and more popular. Simulation has been used in the physical sciences, as well as the engineering sciences, e.g., in aeronautical design, electronic circuit design and mechanical design. The author himself has considerable experience with computer experience with computer simulation. Simulation has generally proven to be cost-saving, time-saving and effective. In engineering applications, e.g., simulation has made it unnecessary to produce or manufacture the prototypes for any new product designs for the purpose of feasibility studies, which could be costly affairs. The feasibility studies are now carried out directly through the simulation exercises, whereby it is possible to quickly find out whether the designs would work or not. The only serious obstacle to simulation appears to be the cost of the simulation software itself, which could come up to many thousands of dollars, so that a cost/benefit analysis for using the software is necessary. The benefits and cost-savings for using the software should outweigh the cost of the software itself in order for its use to be justifiable. The other problem might be a technical one; some simulation software are complex and not user-friendly, requiring a long learning-curve, and these are usually the more powerful software with more built-in features. Nevertheless, once these software have been mastered their use would bring about relatively greater and more effective results. Some might feel that such powerful, and hence more expensive, software are an overkill and prefer to go for something at the lower end which is also cheaper. Such software usually have powerful features such as allowing one to have "walk-through" views and "inside-out" views of an object which would be physically impossible otherwise. Such are the great powers of simulation software now. The simulation of turbulence with powerful software and computers would certainly prove to be very useful.

As nonlinear equations such as the Navier-Stokes equations have to rely on approximation and exact solutions are highly unlikely, especially for turbulent fluid movement, a statistical guide based on actual data collected such as the guides or tables described just above is definitely a boon. A statistical guide that is based on reliable data collected and which has been put to the test and fine-tuned would be in a better position to lead us to more accurate approximations than the Navier-Stokes equations.

From the data in the above-mentioned statistical tables it is now possible, e.g., to make approximations or estimates of various fluid velocities (as well as the directions of various fluid motions) with various probabilities of occurrence for the various Reynolds numbers, such as the following:-

1) x metres/second (accompanied by the appropriate "velocity" diagram): a percent probability of occurrence
2) y metres/second (accompanied by the appropriate "velocity" diagram): b percent probability of occurrence
3) z metres/second (accompanied by the appropriate "velocity" diagram): c percent probability of occurrence

Etc.

The Navier-Stokes equations do not have any allowance for such probabilistic approximations. These partial differential equations have been found to be solvable for the two-dimensional case, i.e., for each of the equations a function (or, solution) could be found that, when substituted for the dependent variable in the equation, leads to an identity. But for three dimensions finding such functions (or, solutions) has been a problem. In other words, though the formulas could be found to describe a two-dimensional fluid motion, such formulas are not available or obtainable for the three-dimensional case, e.g., the case for turbulent fluid motion. Evidently, the geometry of a three-dimensional fluid motion is rather complex, while finding the formulas to describe this complex three-dimensional movement of fluid is a difficult task. To understand this difficulty, consider the motion of a speck of particle carried along in a flowing fluid. The speck could be thrust first in one direction, then another, and another, and so on, sometimes moving in a fairly straight line, other times spiraling around as the current takes it along. This movement of the speck is also the movement of the fluid which carries it along. It is a three-dimensional movement that appears chaotic or turbulent - it is rather complicated and, therefore, not amenable to a description by some formulas. The Navier-Stokes equation makes use of differentiation to obtain the rate of change of some changing quantity, and, in order to do this the value or position or path of that quantity has to be given by an appropriate formula. Differentiation then acts upon this formula to produce another formula which gives the rate of change. Since the formulas for the three-dimensional case, e.g., turbulent fluid motion, are unobtainable the only recourse is approximation. Is there any hope of discovering these formulas in the future? The author is much pessimistic. Let us look again at the case of turbulent fluid motion, an essentially three-dimensional phenomenon. As stated earlier, a curve describing a fluid's movement under conditions of turbulence could be expected to be rough and nonlinear thus making it not possible to derive the differential equations (or formulas) for describing this curve, making predictions relating to the fluid's movement very difficult if not impossible. Moreover, turbulence or chaos implies disorder, irregularity, lack of discernable pattern, confusion and puzzlement. This implies that turbulence could never be described by formulas or differential equations, and, to be able to obtain the formulas or differential equations for turbulence and thus be able to make predictions relating to turbulence would mean that the so-called turbulence is not really turbulence at all, a contradiction. (Do the editors of our dictionaries then have to revise the meanings for turbulence and chaos? According to the Encyclopedic World Dictionary published by Paul Hamlyn, "turbulence" could be defined as "the haphazard secondary motion due to eddies within a moving fluid", or, "irregular motion of the atmosphere, as that indicated by gusts and lulls in the wind", "turbulent flow" is "fluid flow in which the motion at any point varies rapidly in magnitude and direction", and, "chaos" could be defined as "utter confusion or disorder, wholly without organisation or order", or, "the infinity of space or formless matter supposed to have preceded the existence of the ordered universe".) It is thus absurd for us to expect the Navier-Stokes equations to have the solution for a three-dimensional phenomenon such as turbulent or chaotic fluid motion. It is hence appropriate to have another mathematical technique to deal with this difficult situation instead, viz., the statistical method, taking into consideration the example of statistical mechanics in quantum theory. As the outcomes of turbulence or chaos could never be predicted with certainty and the predictions might not even be highly accurate, it is logical and practical to consider the possible outcomes of this phenomenon in a statistical or probabilistic way. Much attempts have already been carried out to understand turbulence or chaos, which is now a hot subject. The author thinks that even if we understand the causes and mechanics or physics of turbulence or chaos it would be naïve to believe it is possible to obtain the formulas for describing such a phenomenon, for, to say that we know how an object which causes confusion and puzzlement (a chaotic object) would behave is to say that we are not confused and puzzled by this object, a self-contradiction. As described above, the logistic equation has been a popular model for chaos but, here again, it would be naïve, and, self-contradicting, to believe that the logistic equation is a sufficient formula for making predictions relating to chaos. Some others might state that only when we could not predict the outcomes of a phenomenon could we regard that phenomenon as chaotic but once we could get those predictions the phenomenon is no more chaotic, which would imply that chaos is transitional and subjective. These are possibly the ones who believe that there is a solution for the Navier-Stokes equations in the three-dimensional case. Simply put, if chaos is predictable it is not chaos. Only when it is really unpredictable could it be chaos. Mathematically, and, objectively speaking too, it has never been found possible to derive the differential equations to describe a nonlinear phenomenon such as turbulence, as there is no regular pattern found in turbulence (which is fluid flow in which the motion at any point varies rapidly in magnitude and direction), and, mathematics, which, in a sense, is the study and analysis of patterns, simply found nothing possible to study or analyse in turbulence since it does not display any discernable, set pattern or regularity, except for the presence of eddies, ripples and whorls, which in the terms of fractal geometry could be described as a fractal characteristic. The only plausible solution appears to be a statistical one, wherein there is some hope of discovering some meaningful patterns or orderly features when large samples of data are analysed. (According to the precepts of fractal geometry, phenomena which appear random when viewed en masse display some orderliness and pattern which could be regarded as a fractal characteristic.) For example, the prime numbers are very random and haphazard entities, yet, when viewed en masse they display a regularity in the way they thin out, whereby it is affirmed that the number of primes not exceeding a given natural number n is approximately $n/\ln n$, in the sense that the ratio of the number of such primes to $n/\ln n$ eventually approaches 1 as n becomes larger and larger, $\ln n$ being the natural logarithm (to the base e) of n (vide the prime number theorem proved in 1896 by Hadamard and Vallee-Poussin).

With statistical methods, such as that described above, it is now possible to evaluate turbulence on a probabilistic basis, which is a more practical and realistic way of looking at turbulence, whose outcomes are uncertain, irregular, haphazard and very difficult if not impossible to predict. As the Navier-Stokes equations fare very poorly in the three-dimensional case, i.e., in the case of turbulence, a new mathematical technique for making approximations for the three-dimensional case, such as the statistical one described above, should be given a chance to take over, to supplant the Navier-Stokes equations, or, at least, to complement them.

After all, the results of the differential equations, such as the Navier-Stokes equations, would still have to be confirmed by actual physical data, or, physical experiment. The more direct, faster or more efficient method of understanding and making forecasts pertaining to the various levels of turbulence or chaos is evidently to execute a well-planned computer simulation exercise (or, at least, a well-planned physical experiment) and apply the proper statistical technique in the interpretation of the data which are obtained through the computer simulation exercise. Such a procedure has been described in detail above.

We should be realistic and refrain from expecting certainly true or almost certainly true forecasts from our "mathematisation" of turbulence or chaos. It is more reasonable or realistic to expect forecasts with only some degree of probability of being true, and, where the forecasts do indeed turn out to be totally true on occasions they should be regarded as rare phenomena not unlike the cases of some fortunate persons hitting the lotteries. In fact, if the day ever arrives when an equation is available for accurately forecasting the outcome of turbulence or chaos, turbulence or chaos would be obsolete. With the lack of a discernable pattern or patterns in turbulence or chaos, such an equation evidently would never be found. In other words, turbulence or chaos, due to its excessively nonlinear, excessively random or patternless, nature, would be practically impossible to describe geometrically through any mathematical equation, e.g., the Navier-Stokes differential equation. A sort of fixed, geometrical shape or pattern has to be discernable to be describable by a mathematical equation. However, in the case of turbulence or chaos, the object in turbulence evidently takes on "different geometrical shapes all the time". Differential equations such as the Navier-Stokes equation are thus bound to fail in helping us to understand turbulence. On the other hand, statistical methods evidently have more chance of success; no matter how slim the chance of success with statistical methods might be, it is evidently still better than that of other mathematical methods. It should be remembered that in the case of statistical methods, the greater the sample size utilised is the greater the chance of success would be. They are therefore the best hope for success for understanding turbulence.

APPENDIX

The Self-Similarity Concept And Fractal Geometry

The formulation of the self-similarity concept has brought fame to Mitchell Feigenbaum, who had worked in the Los Alamos Laboratory in the early 1970s. This concept, upon which the method of renormalisation in perturbation theory is based, postulates that there is a tendency of identical mathematical structures to recur on many levels. Within a given structure, there would be smaller copies of the same structure, their sizes being determined by the scaling factor. Feigenbaum found that at the utmost tips of the fig-tree, there is some mathematical structure which remains the same when its size is changed (enlarged) by a scaling factor of 4.669, which is found to be a constant like pi (3.142). This structure is the shape of the fig-tree itself. In other words, little whorls could be found within big whorls. Renormalisation has been a well-established technique in chaos theory/fractal geometry and is a mathematical trick which functions rather like a microscope, zooming in on the self-similar structure, removing any approximations, and filtering out everything else. All this shows the universality of some features of chaos. That is, some kind of order or pattern could be found in or is inherent in disorder, turbulence or chaos.

In Feigenbaum's famous fig-tree example, for instance, there is a self-similar mathematical pattern or structure (which is the shape of the fig-tree itself) in the various parts of the fig-tree, i.e., its trunk to bough section, bough to branch section, branch to twig section and twig to twiglet section. Such self-similar mathematical pattern or structure, or, fractal characteristic, could also be found in other aspects of nature, e.g., waves, turbulence or chaos, the structures of viruses and bacteria, polymers and ceramic materials, the universe and many others, even the movements of prices in financial markets, the growths of populations, the sound of music, the flow of blood through our circulatory system, the behaviour of people en masse, etc., which have all spawned a relatively new and important branch of mathematics with wide practical applications known as fractal geometry, which has been pioneered by Benoit Mandelbrot. As a matter of fact, self-similarity or fractal characteristic could be regarded as the fundamental mathematical aspect found in practically everything in nature, and, this new branch of mathematics, fractal geometry, besides having a great practical impact on us also gives us a more profound vision of the universe in which we live and our place in it.

18 A PROOF OF THE RIEMANN HYPOTHESIS

The Riemann hypothesis is an important outstanding problem in number theory as its validity will affirm the manner of the distribution of the prime numbers. It posits that all the non-trivial zeros of the zeta function ζ lie on the critical strip between $\mathrm{Re}(s) = 0$ and $\mathrm{Re}(s) = 1$ at the critical line $\mathrm{Re}(s) = 1/2$. The important question is whether there would be zeros appearing at other locations on this critical strip, e.g., at $\mathrm{Re}(s) = 1/4, 1/3, 3/4$, or, $4/5$, etc., which would disprove the Riemann hypothesis. This chapter provides an indirect proof or proof by contradiction (*reductio ad absurdum*) of the Riemann hypothesis.

According to the precepts of fractal geometry, phenomena which appear random when viewed en masse display some orderliness and pattern which could be regarded as a fractal characteristic. For instance, the prime numbers are very random and haphazard entities, yet, when viewed en masse they display a regularity in the way they thin out, whereby it is affirmed that the number of primes not exceeding a given natural number n is approximately $n/\log n$, in the sense that the ratio of the number of such primes to $n/\log n$ eventually approaches 1 as n becomes larger and larger, $\log n$ being the natural logarithm (to the base e) of n (vide the prime number theorem proved in 1896 by Hadamard and de la Vallee-Poussin). In other words, the prime number theorem, which is the direct outcome of the Riemann hypothesis, states that the limit of the quotient of the 2 functions $\pi(n)$ and $n/\log n$ as n approaches infinity is 1, which is expressed by the formula:

$$\lim_{n \to \infty} \pi(n)/(n/\log n) = 1 \qquad (1)$$

the larger the number n is, the better is the approximation of the quantity of primes, as is implied by the above formula where $\pi(n)$ is the prime counting function (π here is not the π which is the constant 3.142 used to compute perimeters and areas of circles, but is only a convenient symbol adopted to denote the prime counting function)

All this is in spite of the fact that the primes are scarcer and scarcer as n is larger and larger.

The prime number theorem could in fact be regarded as a weaker version of the Riemann hypothesis which posits that all the non-trivial zeros of the zeta function ζ on the critical strip bounded by $\mathrm{Re}(s) = 0$ and $\mathrm{Re}(s) = 1$ would be at the critical line $\mathrm{Re}(s) = 1/2$. For a better understanding of the close connection between the prime number theorem and the Riemann hypothesis, it should be noted that Hadamard and de la Vallee Poussin had in 1896 independently proven that none of the non-trivial zeros lie on the very edge of the critical strip, on the lines $\mathrm{Re}(s) = 0$ or $\mathrm{Re}(s) = 1$ - this was enough for deducing the prime number theorem. The locations of these non-trivial zeros on the critical strip could be described by a complex number $1/2 + bi$ where the real part is $1/2$ and i represents the square root of -1. It had already been proven that there is an infinitude of non-trivial zeros at the critical line $Re(s) = 1/2$ on the critical strip between $Re(s) = 0$ and $Re(s) = 1$. The moot question is whether there would be any zeros off the critical line $Re(s) = 1/2$ on the critical strip between $Re(s) = 0$ and $Re(s) = 1$, e.g., at $Re(s) = 1/4$, $1/3$, $3/4$, or, $4/5$, etc., the presence of any of which would disprove the Riemann hypothesis. So far, no such "off-the-critical-line" zeros has been found.

The validity of the Riemann hypothesis would evidently imply the validity of the prime number theorem (which as described above is the offspring and weaker version of the Riemann hypothesis) though the validity of the prime number theorem does not imply the former. Nevertheless, both of them have one thing in common in that they are both concerned with the estimate of the quantity of primes less than a given number, with the Riemann hypothesis positing a more exact estimate of the quantity of primes less than a given number. But, on the other hand, what would be the result if the Riemann hypothesis were false? We will come back to this later.

Meanwhile, more about the non-trivial zeros of the zeta function $\zeta(s)$ defined by a power series shown below:

$$\zeta(s) = \sum_{n=1} 1/n^s = 1 + 1/2^s + 1/3^s + 1/4^s + 1/5^s + \ldots \qquad (2)$$

At the critical line $Re(s) = 1/2$ on the critical strip between $Re(s) = 0$ and $Re(s) = 1$ all the non-trivial zeros would be found on an oscillatory sine-like wave which oscillates in spirals, there being an infinitude of these spirals (representing the so-called complex plane). All the properties of the prime counting function $\pi(n)$ are in some way coded in the properties of the zeta function ζ, evidently resulting in the primes and the non-trivial zeros being some sort of mirror images of one another - the regularity in the way the primes progressively thin out and the progressively better approximation of the quantity of primes towards infinity by the prime counting function $\pi(n)$ mirror or reflect the regularity in the way the non-trivial zeros of the zeta function ζ line up at the critical line $Re(s) = 1/2$ on the critical strip between $Re(s) = 0$ and $Re(s) = 1$, the non-trivial zeros becoming progressively closer together there, with no zeros appearing anywhere else on the critical strip, and, all this has been found to be true for the 1st. 10^{13} non-trivial zeros.

Riemann had posited that the margin of error in the estimate of the quantity of primes less than a given number with the prime counting function $\pi(n)$ could be eliminated by utilizing the following J function which is a step function involving the non-trivial zeros expressed in terms of the zeta function ζ, which has been shown to be effective (2 steps are involved here - first, the prime counting function $\pi(n)$ is expressed in terms of the $J(n)$ function, then the $J(n)$ function is expressed in terms of the zeta function ζ, with the $J(n)$ function forming the link between the counting of the prime counting function $\pi(n)$ and the measuring (involving analysis and calculus) of the zeta function ζ, which would result in the properties of the prime counting function $\pi(n)$ somehow encoded in the properties of the zeta function ζ):

$$J(n) = Li(n) - \sum_{p} Li(n^{p}) - \log 2 + \int_{n}^{\infty} dt/(t(t^2 - 1) \log t) \qquad (3)$$

where the 1st. term $Li(n)$ is generally referred to as the "principal term" and the 2nd. term $\sum_{p} Li(n^{p})$ had been called the "periodic terms" by Riemann, Li being the logarithmic integral

The above formula may look intimidating but is actually not. The 3rd. term log 2 is a number which is 0.69314718055994... while the 4th. term $1/(t(t^2 - 1) \log t)$ which is an integral representing the area under the curve of a certain function from the argument all the way out to infinity can only have a maximum value of 0.1400101011432869.... Since these 2 terms taken together (and minding the signs) are limited to the range from -0.6931... to -0.5531..., and since the prime counting function $\pi(n)$ deals with really large quantities up to millions and trillions they are much inconsequential and can be safely ignored. The 1st. term or principal term $Li(n)$, where n is a real number, should also be not much of a problem as its value can be obtained from a book of mathematical tables or computed by some math software package such as *Mathematica* or *Maple*. However, special attention should be given to the 2nd. term $\sum_{p} Li(n^{p})$

which concerns the sum of the non-trivial zeros of the zeta function ζ (p in this 2nd. term is a "rho", which is the 17th. letter of the Greek alphabet, and it means "root" - a root is a non-trivial zero of the Riemann zeta function ζ - a root here is a solution or value of an unknown of an equation which could be factorized). Riemann had evidently called the 2nd. term "periodic terms" as the components there vary irregularly.

The prime number theorem asserts that $\pi(n) \sim Li(n)$ (technically $Li(n) = \int_{2}^{n} dx/\log (x)$) which also implies

the weaker result that $\pi(n) \sim n/\log n$. However, with $Li(n)$ the prime count estimate would have a margin of error. The Riemann hypothesis asserts that the difference between the true number of primes $p(n)$ and the estimated number of primes $q(n)$ would be not much larger than $\sqrt{n}$. With the above $J(n)$ function we could eliminate this margin of error and obtain an exact estimate of the quantity of primes less than a given number:

$J(n)$ = exact quantity of primes less than a given number

Since the 3rd. and 4th. terms of the $J(n)$ function are inconsequential and can be safely ignored, as is described above, deducting the 2nd. term from the 1st. term should be sufficient:

$$J(n) = Li(n) - \sum_{p} Li(n^p) = \text{exact quantity of primes less than a given number}$$

The above in a nutshell shows the intimate relationship between the primes and the non-trivial zeros of the zeta function ζ, the primes and the non-trivial zeros being some sort of mirror images of one another as is described above, with the distribution of the non-trivial zeros being regarded as the music of the primes by mathematicians.

We return to the question of the consequence of the falsity of the Riemann hypothesis. Let us here assume that the Riemann hypothesis is false, i.e., there are also zeros found off the critical line $\text{Re}(s) = 1/2$ on the critical strip between $\text{Re}(s) = 0$ and $\text{Re}(s) = 1$, e.g., at $\text{Re}(s) = 1/4$, $1/3$, $3/4$, or, $4/5$, etc., and see the consequence. What would be the significant implication of this assumption? The falsity of the Riemann hypothesis would imply that the distribution of the zeros of the zeta function ζ on the critical strip between $\text{Re}(s) = 0$ and $\text{Re}(s) = 1$ has lost the regularity of pattern which is characteristic of the non-trivial zeros at the critical line $\text{Re}(s) = 1/2$ and which is described above, and is now disorderly and irregular. This would in turn imply that the distribution of the primes is also similarly disorderly and irregular since the primes and the non-trivial zeros of the zeta function ζ are intimately linked and are some sort of mirror images of one another - any changes in one of them would be reflected in the other on account of their intimate link - note that the zeta function ζ has the property of prime sieving (compare: sieve of Eratosthenes) encoded within it, the properties of the prime counting function $\pi(n)$ being somehow encoded in the properties of the zeta function ζ, so that if the zeros generated were disorderly and irregular it would mean that the distribution of the primes were also similarly disorderly and irregular - the characteristic of the primes on the input side of the function determines the characteristic of the zeros on the output side of the function (i.e., the distribution of the primes determines the distribution of the zeros, so that from a study of the distribution of the zeros the distribution of the primes could be deduced and vice versa), which is expected for a function. The overall result would be that the more orderly the distribution of the zeros is the more orderly would be the corresponding distribution of the primes, the more disorderly the distribution of the zeros is the more disorderly would be the corresponding distribution of the primes, and, vice versa. But, according to the prime number theorem, or, prime counting function $\pi(n)$, which is closely connected with the Riemann hypothesis itself being an offspring and weaker version of it as is described above, there is instead actually a regularity in the way the primes thin out, with the prime counting function $\pi(n)$ even providing a progressively better estimate of the quantity of primes towards infinity - this progressively better estimate would not be possible if the primes behave really badly and are really highly disorderly and irregular - there is no such really great disorder or irregularity among the primes, a state of affair which is evidently affirmed by the fact that the corresponding non-trivial zeros at the critical line Re(s) = 1/2 on the critical strip between Re(s) = 0 and Re(s) = 1 display regularity in the way they line up at the critical line Re(s) = 1/2, the non-trivial zeros becoming progressively closer together there with no zeros appearing anywhere else on the critical strip (all of which has been found to be true for the 1st. 10^{13} non-trivial zeros - an important point to note is that though the non-trivial zeros at the critical line Re(s) = 1/2 become more and more closely packed together the farther along we move up this critical line while the primes occur farther and farther along the number line, the density of the one is approximately the reciprocal of the density of the other wherein the complementariness, regularity, symmetry is evident), this regularity of the distribution of the non-trivial zeros mirroring the regularity of the distribution of the primes as is explained above. Our assumption of the falsity of the Riemann hypothesis has thus resulted in a contradiction of the actual distribution of the primes and the actual distribution of the corresponding non-trivial zeros at the critical line Re(s) = 1/2 on the critical strip between Re(s) = 0 and Re(s) = 1. If our assumption that the Riemann hypothesis is false is correct, the prime number theorem would be false as there would be great disorder and irregularity among the primes with no regularity in the way the primes thin out and without the prime counting function $\pi(n)$ providing a progressively better estimate of the quantity of primes towards infinity (this progressively better estimate of the quantity of primes actually implies some regularity in the distribution of the primes). However, as is explained just above the prime number theorem is not false; it had in fact been proven through both non-elementary methods (by Hadamard and de la Vallee Poussin) and elementary methods (by Erdos and Selberg later) and is indubitably true. Therefore, our assumption of the falsehood of the Riemann hypothesis is at fault. The Riemann hypothesis cannot be false and has to be true.

APPENDIX

THE RIEMANN ZETA FUNCTION AND THE PRIME NUMBERS
The Riemann zeta function $\zeta(s)$, shown below, is the sum over all natural numbers n:

$$\zeta(s) = \sum_{n=1} 1/n^s = 1 + 1/2^s + 1/3^s + 1/4^s + 1/5^s + \ldots$$

The function could also be written in the following way (using Euler's product formula) showing its connection with the prime numbers:

$$\zeta(s) = \prod_{p \text{ prime}} p^s/p^s - 1 = 2^s/2^s - 1 \times 3^s/3^s - 1 \times 5^s/5^s - 1 \times 7^s/7^s - 1 \times \ldots$$

where the product is over the consecutive prime numbers p, providing the first hint that the Riemann zeta function $\zeta(s)$ is closely linked to the prime numbers.

19 CONCLUSION

Research mathematics evidently concerns the proving of theorems, e.g., the infinitude of the non-trivial zeros of the zeta function of the Riemann hypothesis, the infinity of the solutions for the Birch-Swinnerton-Dyer conjecture, et al. A proof might just require several lines of text and symbols, or, more, and, in the more extreme case, even hundreds or more pages of text and symbols. But is a proof of a theorem in fact a tautology, i.e., just another way of expressing or stating a theorem so that the theorem is clearly understood and accepted as valid in its logic? That is, is a proof of a theorem in fact the following tautological statement?:

Theorem A = confirmed (or proven) Theorem A

A proof has to be rigorous, water-tight, so it is said, to have no gaps or occasions for doubt. However, the concept of "rigorous" is not clear and difficult to explain. Even the meaning of "rigor" found in a standard dictionary is vague, e.g., stating the meaning as "strictness, severity, or harshness", whatever that really means.

A proof involves reasoning with axioms and/or lemmas. For example, a mathematician might use axioms and/or lemmas to prove Theorem A. After a possibly long chain of reasoning, he might (or might not, sometimes) confirm that Theorem A is really a theorem. The chain of mathematical reasoning in the proof is there to convince its reader of the truth or validity of Theorem A. The mathematical reader however might not be convinced due to lack of understanding or intelligence, bias (e.g., cultural bias which is a common phenomenon), prejudice and what-not; he might even stubbornly not be convinced (for some reason best known to himself). Reasoning or logic should be objective and clear-cut, one might think, but sadly it often appears not which possibly explains all the different schools of thought.

On the other hand, an ultra-intelligent person, possibly an extra-terrestrial, might know that Theorem A is really Theorem A without needing to be convinced by the above-said chain of reasoning, possibly having his own way of ascertaining it, and even possibly by pure, powerful intuition.

Could a proof be non-tautological? One might wonder. A proof would not be tautological if, e.g., the chain of reasoning leads from Theorem A to Theorem B, or, more other theorems:

Theorem A ⟶ Theorem B (et al.)

Of course this also happens when smaller theorems or lemmas help the mathematician to arrive at the main theorem (the so-called tautology).

This kind of reasoning is known as deduction, the kind of reasoning the iconic fictional private detective Sherlock Holmes had famously used. Just from some clues Sherlock Holmes was able to deduce who the criminal or murderer was:

clues $\longrightarrow$ criminal/murderer

Should and could mathematics not also make use of such non-tautological reasoning? In other words, shouldn't mathematical reasoning also be non-tautological, which could result in more theorems being deduced or discovered, as is stated above?

BIBLIOGRAPHY

[1] D. Burton, 1980, Elementary Number Theory, Allyn & Bacon

[2] R. Courant and H. Robbins, revised by I. Stewart, 1996, What Is Mathematics? An Elementary Approach to Ideas and Methods, Oxford University Press

[3] G. H. Hardy and E. M. Wright, 1979, An Introduction To Theory Of Numbers, Oxford, England: Clarendon Press

[4] M. E. Lines, 1986, A Number For Your Thoughts, Adam Hilger

[5] B. B. Mandelbrot, 1977, The Fractal Geometry Of Nature, W. H. Freeman

[6] R. McWeeny, 2002, Symmetry: An Introduction To Group Theory And Its Applications, Dover